TIME-OF-FLIGHT MASS SPECTROMETRY

TIME-OF-FLIGHT MASS SPECTROMETRY

Edited by

D. PRICE
B.Sc., Ph.D., A.R.I.C.
University of Salford

and

J. E. WILLIAMS
Fieldtech. Ltd., London

Based on the Proceedings of the First European Symposium on Time-of-flight Mass Spectrometry held at the University of Salford, July 3–5, 1967

THE QUEEN'S AWARD TO INDUSTRY 1966

PERGAMON PRESS

OXFORD · LONDON · EDINBURGH · NEW YORK
TORONTO · SYDNEY · PARIS · BRAUNSCHWEIG

Pergamon Press Ltd., Headington Hill Hall, Oxford
4 & 5 Fitzroy Square, London W.1
Pergamon Press (Scotland) Ltd., 2 & 3 Teviot Place, Edinburgh 1
Pergamon Press Inc., Maxwell House, Fairview Park, Elmsford, New York 10532
Pergamon of Canada Ltd., 207 Queen's Quay West, Toronto 1
Pergamon Press (Aust.) Pty. Ltd., 19a Boundary Street, Rushcutters Bay, N.S.W. 2011, Australia
Pergamon Press S.A.R.L., 24 rue des Écoles, Paris 5e
Vieweg & Sohn GmbH, Burgplatz 1, Braunschweig

First edition 1969

Library of Congress Catalog Card No. 79-80074

PRINTED IN HUNGARY

08 013444 0

CONTENTS

DELEGATES AND LECTURERS

UNIVERSITY OF SALFORD FIRST EUROPEAN SYMPOSIUM: THE TIME-OF-FLIGHT MASS SPECTROMETER

DELEGATES

DR. R. E. BUHLER, ETH, Zürich, Switzerland.
DR. E. S. LIMONGELLI, Venturimetro, Milan, Italy.
MR. C. RUBELT, University of Saarlandes, Hombourg, Germany.
MR. A. TORRENT, Rhone–Poulenc, Cg St Fons, France.
DR. V. PIACENTE, University of Rome, Italy.
DR. J. ROSENBAUM, Mars Ltd., Slough, England.
DR. G. D'EGIDIO, University of Rome, Italy.
DR. S. TORLASCHI, Montecatini Edisons SPA, Milan, Italy.
MR. S. E. LINDQUIST, University of Uppsala, Sweden.
MR. L. FORNSTEDT, University of Uppsala, Sweden.
MR. V. N. VASILYEV, IAEA, Vienna, Austria.
MR. T. R. KEMP, AEI Ltd., Manchester, England.
MR. M. J. DAY, University of Leeds, England.
MR. E. PRITCHARD, BCURA, Leatherhead, England.
MR. D. WINDSOR, Chemical Inspectorate, Woolwich, England.
DR. P. A. ROBINSON, University of Cambridge, England.
MR. J. M. B. BAKKER, Shell Research Ltd., Sittingbourne, England.
DR. P. F. KNEWSTUBB, University of Cambridge, England.
MR. A. CUTTING, Elliot Tubes Ltd., Borehamwood, England.
MR. J. DELWICHE, University of Liège, Belgium.
MR. H. DORING, Neumuller & Co., Munich, Germany.
DR. J. D. MADELEY, University of Salford, England.
DR. T. HENSHALL, University of Salford, England.
MR. P. CLOUGH, University of Salford, England.
DR. D. DOLLIMORE, University of Salford, England.
MR. P. SPOONER, University of Salford, England.
MR. S. APPLEBY, University of Salford, England.
MR. G. B. NUTT, Shell Research Ltd., Thornton, England.

LECTURERS

DR. D. PRICE, University of Salford, England.
MR. K. JOY, BCURA, Leatherhead England.
MR. K. O. DYSON, University of Cambridge, England.
DR. R. G. DIXON-LEWIS, University of Leeds, England.
DR. R. BOWLES, CERL, Leatherhead, England.
MR. C. D'OYLY-WATKINS, Chemical Inspectorate, Woolwich, England.
DR. J. C. J. THYNNE, University of Edinburgh, Scotland.
DR. D. W. THOMAS, University of Birmingham, England.
MR. D. C. DAMOTH, Bendix Corporation, Cincinatti, U.S.A.
MR. E. J. YOUNGINGER, Bendix Corporation, Cincinatti, U.S.A.
MR. P. WEINREICH, Bendix International, Frankfurt, Germany.
MR. J. E. WILLIAMS, Fieldtech Ltd., London, England.
MR. B. C. MARTIN, Fieldtech Ltd., London, England.
MR. B. D. EDKINS, Fieldtech Ltd., London, England.

PREFACE

MANY books on mass spectrometry are now available. They all deal at length with the theory and applications of the magnetic type of instrument, but little reference is made to time-of-flight operation.

This book is the first to be devoted entirely to the rapidly expanding field of time-of-flight mass spectrometry. This development is due in the main to advantages of the open geometry of the ion source and the high scan rate of the time-of-flight instrument. The chemical problems to which it has been applied include fast reactions, free radical chemistry, ion-molecule reactions, negative ion studies, Knudsen cell measurements, monitoring of gas chromatograph effluents and routine analysis.

There are now about 300 time-of-flight mass spectrometers in use throughout the world. Some fifty of these are in Europe. The First European Symposium was held in order to bring together the users on this side of the Atlantic. The proceedings of the symposium, as published in this book, are intended to fulfil the need for a textbook on time-of-flight mass spectrometry. The book will be of obvious value to the user of a time-of-flight mass spectrometer. It will also enable the non-user to become acquainted with the applications and advantages of the instrument. It is hoped that this book will encourage others to apply the technique to their own particular problems. Thus can the subject of time-of-flight mass spectrometry be enlarged. The book gives detailed consideration to the chemical applications of the time-of-flight mass spectrometer. A bibliography up to June 1968 is included.

The symposium organizers wish to thank The Bendix Corporation, U.S.A., and Fieldtech. Ltd., U.K., for the financial support and the University of Salford for the facilities used during the symposium.

D. P.
J. E. W.

INTRODUCTION

THE MA-1 AND OTHER TIME-OF-FLIGHT MASS SPECTROMETERS

D. C. DAMOTH

Scientific Instruments and Vacuum Division, The Bendix Corporation, U.S.A.

MASS spectrometers are the most sensitive of the general analysis instruments. The Time-of-Flight Mass Spectrometer (TOFMS) has the widest range of application and the maximum speed of analysis of the various types of mass spectrometers.

Unlike the geometric (magnetic) deflection and filter-type mass spectrometers, the TOFMS collects *all* the ion masses in the spectrum on each cycle of the instrument (Fig. 1). Thus, by means of an oscilloscope, every mass peak may be monitored at intervals ranging from 10 to 100 μsec. This ability to see the entire spectrum on the oscilloscope, while recording the spectrum by the analog output, plus the widest range of inlet capabilities, separates the TOFMS from the other types.

Figure 2 is a schematic diagram of the TOFMS ion source, as designed by Mr. W. C. Wiley. A large opening between the backing plate and the first ion grid (IG-1) permits the introduction of liquid, gas or solid samples without obstruction. This line-of-sight sampling allows reactive and unstable species to be introduced directly into the ion source region with little probability of wall collisions.

In the source operation thermionically emitted electrons are held back by an appropriate bias on the first electron grid. At the start of each spectrum the electrons are accelerated through the ionization region by pulsing the grid positive with respect to the filament. The energy with which the electrons pass through the source region is controlled by adjusting the bias on the filament. This bias is usually 0–100 V in respect to the ionizing region

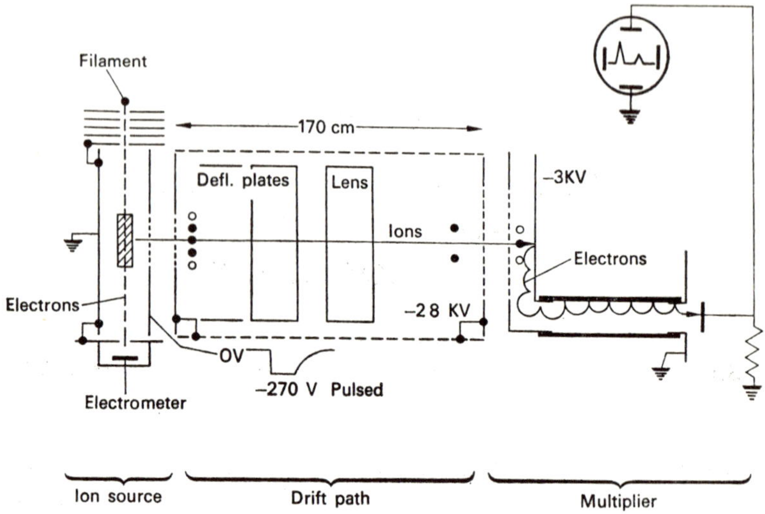

FIG. 1. TOFMS schematic.

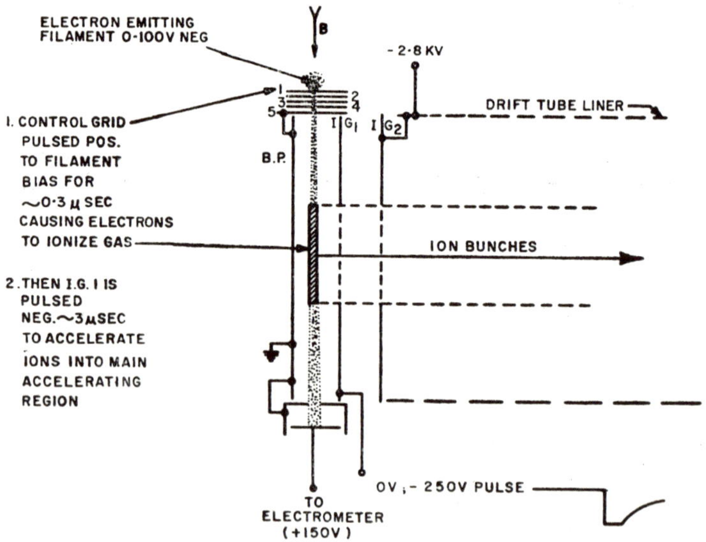

FIG. 2. S-12-101 operation (pulsed ionization).

which is at, or near, ground potential at the time of ionization. An electrode on the opposite side of the source region picks up the electron beam. This current is amplified and a servo loop either adjusts the drive on the filament or changes the control grid bias to keep the intensity of the ionizing beam constant. The electron beam is usually on for from 0·25 μsec to several microseconds. (In one mode of operation the beam may be on continuously, except during the ion drawout period.) The electron beam bombards the sample gas molecules in the source region and ionizes a fraction of them.

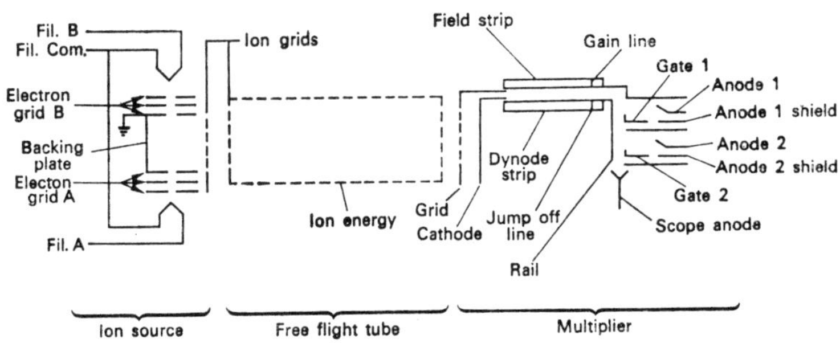

FIG. 3. MA-1 mass analyzer.

For the first time in a TOFMS, the MA-1 employs two filaments, one on either side of the ion source (Fig. 3). The electron gun geometry is symmetrical, so a single switch changes from one filament to the other. If one filament chances to burn out during use, operation can be immediately resumed with the other filament and the defective one replaced in a few minutes at the operator's convenience.

The open configuration of the TOF ion source results in a lower operating temperature which permits analysis of thermally unstable compounds. If the temperature of the source must be raised to facilitate removal of "sticky" compounds, this can be done easily.

After the electron beam is shut off, the ions are accelerated out of the ionization region by the application of a fast rise pulse to the first (drawout) ion grid. Two factors which affect resolution have been determined at this location. First, the rise time of the drawout pulse should be 0·1 μsec (10^{-7} sec) or less. A slower rise time on the drawout pulse will cause a very small amount of spreading of the ions, thereby reducing resolution. Secondly, the drawout grid must be planar to within a few thousandths of an inch. If not, it will cause non-uniform acceleration of the ions which causes loss of resolution by axial spreading of the ion

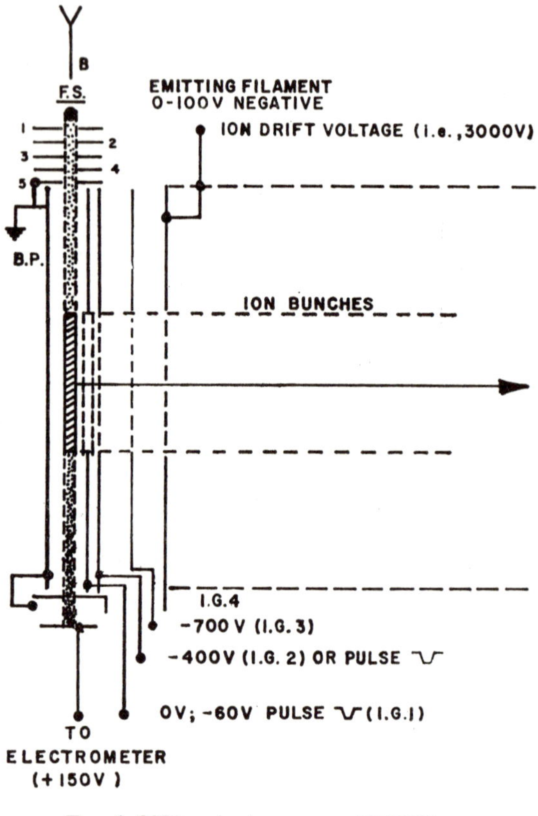

FIG. 4. 3000 series ion source TOFMS.

bunches and loss of sensitivity by radial spreading so that they miss the detector.

The ionization region is essentially a ground potential "box" through which the electron beam passes. Maintaining freedom from field penetration has been accomplished by using a double mesh separated by a 0·020 in. plate for the first grid. This is primarily important for ionization potential work. In the 3012 and 3015 instruments another ion grid is located immediately adjacent to ion grid number one (Fig. 4). This auxiliary grid can be operated either with an adjustable d.c. potential or with a pulse coincident with the pulse on ion grid one. This also helps reduce field penetration.

A second accelerating grid at an intermediate potential is now used before the energy potential grid. This second grid is usually operated at 0·3 of the potential on the third or energy grid. Evidence indicates that focusing is improved with this arrangement.

The spacings in the ion source have been investigated empirically and the distance between the backing plate and first ion grid has been optimized at about 0·120 in. for a 1–2 m drift tube with the electron beam 0·050 in. from the drawout grid. Preliminary investigation with 0·008 in. electron slits indicate a noticeable resolution improvement can be obtained with the narrower electron beam. 0·015 in. slits are customary. The ions are usually accelerated to final velocity within 1 cm of the ionization region, but this does not appear to be critical.

In the field-free drift tube, the ions separate out in time according to their mass-to-charge ratio (m/e) by virtue of their different velocities. (Fig. 1) The lighter bunches speed ahead while the heavier ions lag behind. Deflection plates are used to center the beam on the detector in most instruments. However, if the first drawout grid is indeed flat and orthogonal to the flight tube axis, the deflection plates can usually be turned off during operation. This eliminates the necessity of correcting the flight path and thus causing warped trajectories which reduce resolution and sensitivity.

In fast reaction work it is desirable to have the deflection plates at the same potential as the drift tube liner. With this configuration, the potential on the plates is not perturbed during the reaction process, thus preventing sporadic deflection of the ion beam during the measurement. This seems to be a fundamental factor in enabling Dick Meyer of Sandia to achieve better reproducibility on spectra generated 20 μsec apart. By getting better ion transmission, he found that he could get peaks with as many as 100,000 ions/cycle running at 50 kc repetition rates. In the Model MA-1 Spectrometer, where the distance from the drawout grid to the detector is only 20 cm (8 in.), there are no deflection plates at all (Fig. 3).

In the ion source region a collimating magnetic field is used to focus the electron beam. It has been customary to use a compensating magnet immediately after the ion source to cancel part of the fringing field of the collimating magnet and to readjust the trajectories of the ions toward the detector. Now it has been determined that it is also advisable to use a compensating magnet near the magnetic electron multiplier to adjust for deviations caused by the fringing flux of the multiplier magnet. In this man-

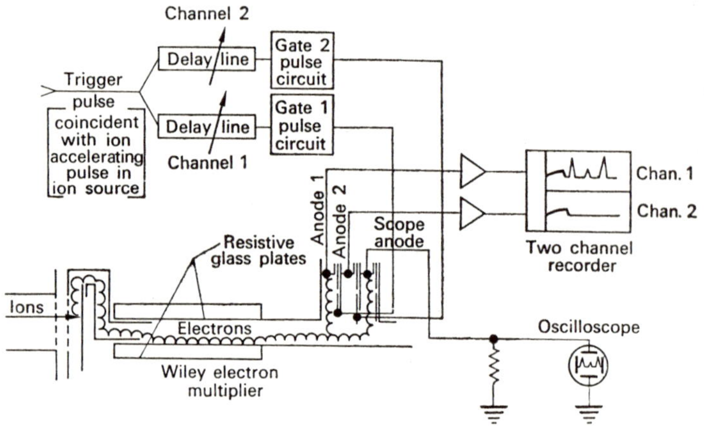

FIG. 5. Schematic of recording output system for Bendix TOFMS.

ner, nearly perfect magnetic compensation of the ion trajectories is achieved.

At the terminus of the flight path of the ions, the separated sheets of ions strike a cathode with approximately 3 kV energy and dislodge electrons at about a 1 to 1 ratio. Crossed electric and magnetic fields cause the electrons to follow cycloidal paths (Fig. 5). By properly adjusting the equipotential lines of the electric field, the electrons can be directed wherever desired because the cusp of the cycloid will try to return to the starting potential. The cycloiding electrons enter the magnetic electron multiplier (MEM) where the multiplication process is continued by successive impacts on the dynode strip. Since each primary electron dislodges approximately five electrons in the first multiplication step at cathode number two, the probability of getting a reduced gain per stage for every electron is remote, thereby improving pulse height distribution. After 40 to 60 impacts on the dynode strip, gains of 10^5 to over 10^7 are achieved.

The resistive glass strip dynodes are unique in themselves. They are made by a process which yields a resistive surface that is very durable. It can be exposed to moisture, and many other atmospheric contaminants, with no adverse affect on the gain characteristics of the glass.

After leaving the multiplying region, the electron sheets are directed into one of the output channels. The final channel is the oscilloscope gate and is always "on". Any electron pulses, not gated into a preceding channel, are picked up by the scope anode and form a series of voltage pulses across a resistor. These voltages are amplified and applied to the vertical plates of the oscilloscope to provide the familiar scope spectrum of the TOFMS.

If one of the earlier gates is on, the electrons are picked up on that channel's anode, amplified, and that signal may be read out on a meter or recorder. The signal to a channel can be integrated to yield the concentration of that mass-to-charge species. The time at which an anode is pulsed can also be varied electrically to give an analog of the mass spectrum. To get an idea of what this means, if one scans with an analog channel over a single mass peak in 1 sec, the time delay applied to that channel has been varied 10,000 times in that second. Thus, literally 10,000 tiny discrete "jumps" are made going over the peak. However, the increments are so small that the normal recorder presents an essentially smooth integration of the shape of the mass peak.

Two additional features on the MA-1 analog as well as on other analogs are the automatic repetitive scanning and logarithmic sensitivity range. These features are finding extensive use in gas chromatograph effluent identification and other qualitative applications. It is particularly advantageous to be able to scan a large dynamic range repetitively with a single recorder channel. The automatic repetitive scan makes it feasible to monitor several peaks over a continuous mass range at short intervals without the tedium of manual switching. Solid state circuitry in the analog amplifiers yields much improved baseline stability for quantitative analysis. Similarly, a new solid state delay circuit invented during the development of the MA-1 yields essentially drift-free mass calibration. This will greatly augment computer data reduction of mass spectra.

A popular development is the "total output integrator" plug-in. This device integrates the entire mass spectrum on every eighth cycle of the TOFMS. (Fig. 6). During the first six cycles the normal analog channels remain on, which allows normal operation with those channels. On the seventh cycle, the oscilloscope is triggered and the spectrum is displayed on the scope. On the eighth cycle, the total output integrator turns itself on and remains

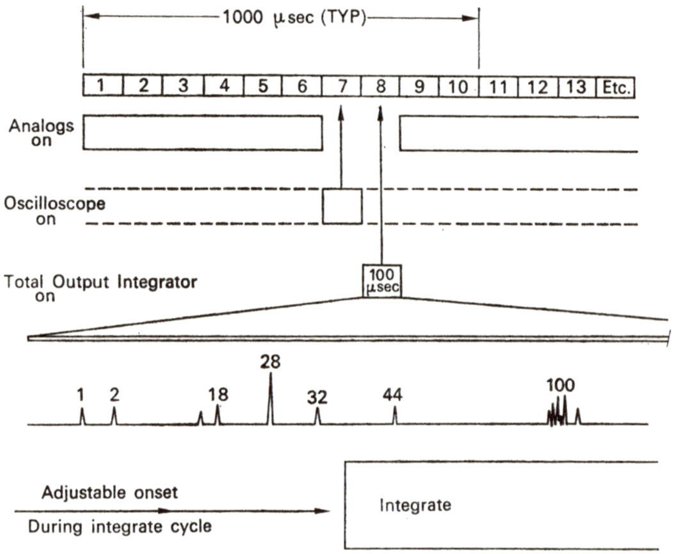

FIG. 6. Total output integrator time sequence.

on until the end of that cycle. This allows integrating the curve due to all the peaks in the spectrum. This sequence is repeated after the eighth cycle.

The device may also be adjusted to integrate only those masses above a chosen mass number. Figure 6 shows the integration starting at about mass 36 or 37. By setting the onset of the integration at an appropriate mass number, helium, carrier gases in gas chromatographs, water, nitrogen, oxygen, etc., can be eliminated from the integration. This technique has been very successfully employed in the food flavor industry in conjuction with GC/MS analysis. A normal capillary flow rate with no split would flood a mass spectrometer, but by operating at a 2 atm pressure differential, with the output into the high vacuum system, satisfactory conditions are maintained. With this method there is less probability of losing very small peaks coming out of the gas chromatograph, or of pyrolyzing them.

Since the mass spectrometer processes must usually take place without significant intermolecular collisions, a vacuum system is required to reduce the gas concentration in the analyzer to a level where particle mean-free paths are long. Figure 7 shows a typical vacuum system for a TOFMS. A working pressure of 10^{-5} torr or lower is obtained with this system. A three-stage mercury

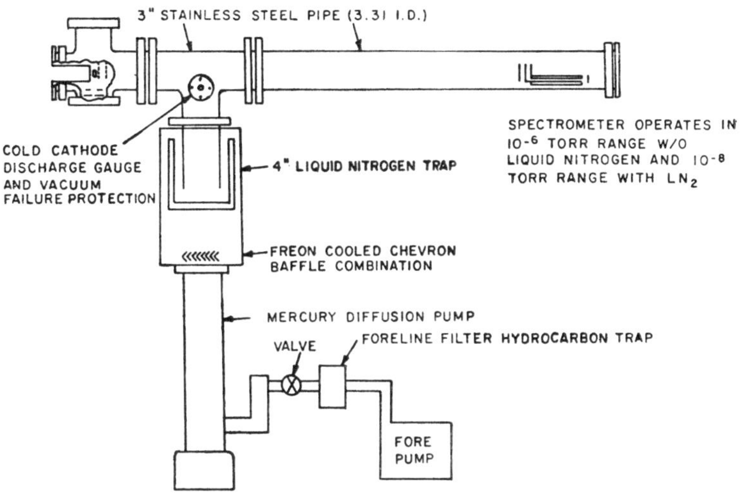

FIG. 7. Typical TOF pumping system.

diffusion pump, a freon cooled chevron baffle and liquid nitrogen trap is the standard pumping system. This design permits continuous operation, without liquid nitrogen, in the 10^{-6} torr range. The addition of liquid nitrogen in the cold trap allows pressures to the 10^{-8} torr range or lower. The mercury diffusion pump is backed up by a forepump with a molecular sieve trap between the diffusion pump and forepump in the larger instruments. This trap prevents hydrocarbon vapors from traveling back into the system and causing a high background. Background reduction is an important factor in obtaining increased sensitivity. The vacuum system employs either a cold cathode discharge gauge or a hot filament Bayard Alpert gauge for approximate pressure measurement and over-pressure protection.

The mercury diffusion pump was chosen by Bendix after careful comparison of available vacuum units. The pumping speed of an ion pump falls off between 10^{-5} and 10^{-3} torr, the region where most work is done with the mass spectrometer. They also become

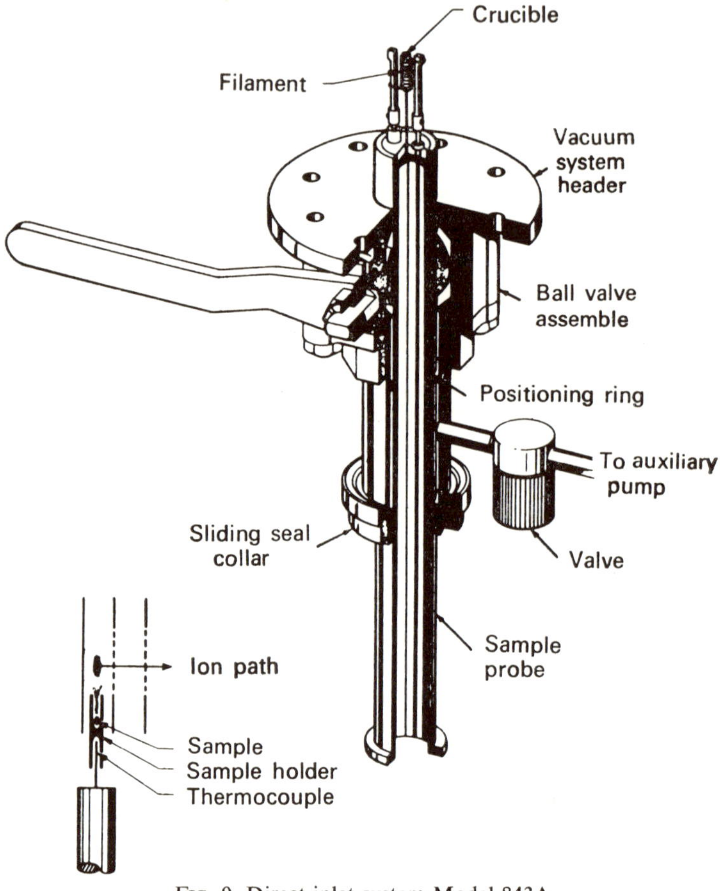

FIG. 9. Direct inlet system Model 843A.

unstable after pumping a gas continuously for a few hours, and will not pump noble gases at all well. Oil diffusion pumps, although less of a problem than mercury pumps in some areas, have the disadvantage that oil contamination in a system is practically impossible to eliminate. Oil decomposition products give a background of practically every mass number in the spectrum.

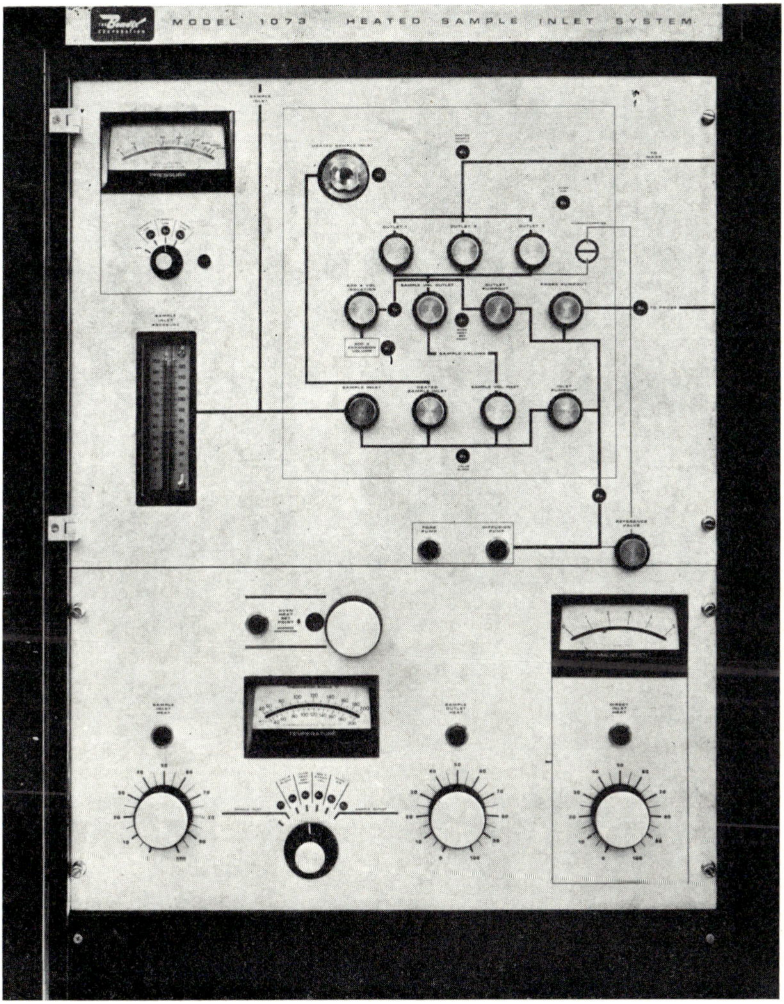

FIG. 8. Model 1073 heated sample inlet system.

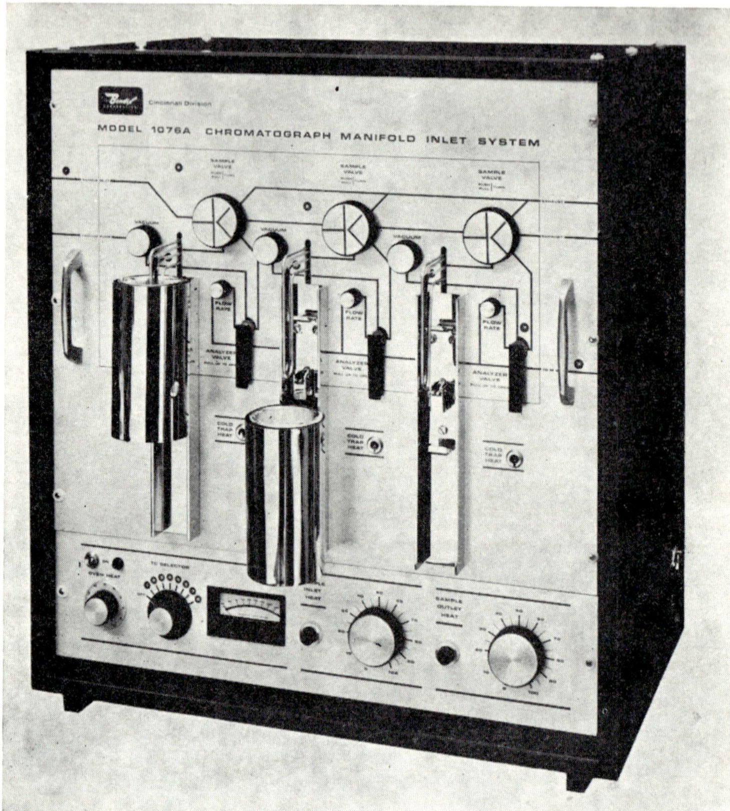

FIG. 10. Model 1076A chromatograph manifold inlet system.

The high efficiency and ease of maintenance of a mercury pumping system allow practically continuous trouble-free operation of the TOFMS.

Several experiments, or apparatus, can be hooked up to the TOF mass spectrometer simultaneously because of the availability of the source region for attachment of accessories.

One of the classical inlets is the heated molecular leak inlet with three molecular leaks in parallel (Fig. 8). The large 5 l. expansion volume can be valved off for very small samples and the approximately 250 ml volume of the manifold used. Obviously, with this system, the sample decay rate will be much faster, but the sensitivity for very small samples is much greater. Also available is the unheated molecular leak direct inlet system for applications where it is not necessary to heat the sample chamber and manifold. The very popular and much copied direct inlet system utilizes a probe and a vacuum lock (Fig. 9). A small crucible surrounded by a coil filament is mounted on top of the probe. The filament heats the crucible and vaporizes a tiny bit of sample directly into the ionization region, providing a maximum sensitivity condition. This permits operation with the sample at minimum temperature, thereby reducing the probability of thermally decomposing the sample. The sample residue remains in the crucible and is taken out when the crucible is removed from the mass spectrometer.

The gas chromatograph manifold system (Fig. 10) basically described by Dr. A. A. Ebert about 1960, has not gained popularity in proportion to its early expectations, primarily due to its high cost. However, the system does represent an ultimate approach to the GC/MS interface problem. Essentially all of the gas chromatograph fraction goes into the mass spectrometer, but the carrier gas is excluded. This is the only link between gas chromatographs and mass spectrometers which allows control over the time domain. With this inlet a particular chosen gas chromatograph peak can be scanned in detail indefinitely. This is important when sorting out some isomers.

The Knudsen cell is a device for examining the vapor species of solid and liquid samples at higher temperatures without significantly disturbing the equilibrium between the gas and the sample (Fig. 11). The sample is elevated to the desired temperature and the molecular beam is projected into the ion source with minimum alteration of the sample species. The molecular beam from the

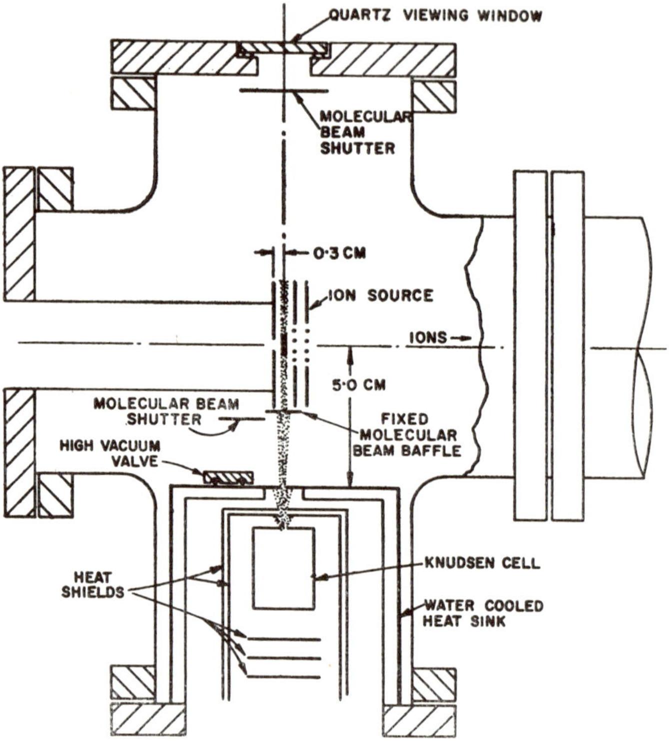

FIG. 11. Knudsen cell operation.

Knudsen cell is orthogonal to the electron beam and the ion trajectory, thus minimizing beam alteration.

The laser microprobe is one of the newest developments (Fig. 12). This is basically a modified direct inlet probe which inserts the sample just below the ionization region of the mass spectrometer. A laser with energy of about 1 joule is focused through appropriate optics to produce a 10–100 μ spot size on the surface of the sample. The position of the laser is continuously adjustable allowing different spots in inhomogeneous samples to be vaporized selectively. This permits some very unusual analysis.

Another approach is to use a high energy laser to achieve temperatures beyond the range of the Knudsen cell studies. This technique is limited to a laser on a TOF instrument because

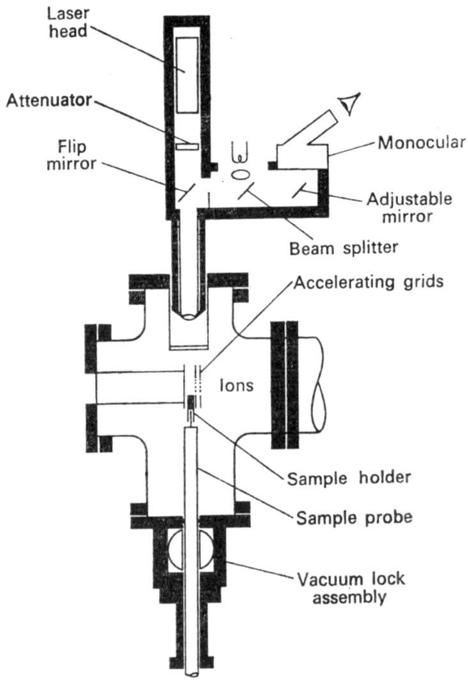

FIG. 12. TOF/laser microprobe.

the sample is vaporized in a few milliseconds. The laser burst itself lasts roughly 0·001 sec. All the species from every cycle must be detected and analyzed in this time. The TOFMS will give from 10 to 100 spectra in the time available and show all of the peaks on each spectrum.

Knox and Vastola at Penn State are doing some fascinating work on a wide range of organic species. Based on their findings, a new approach to organic structure analysis may result. This interesting approach to introducing samples into the TOFMS does not completely cremate the organic sample, but by properly controlling sampling conditions, a very delicate job of "lifting" the organic molecules out of the surface of the sample is accomplished.

The ion–molecule reaction system (Fig. 13) is another unique innovation. It allows direct measurements of pressure in the ionization region by straightforward mechanical means. The TOFMS arrangement of the open source makes possible this

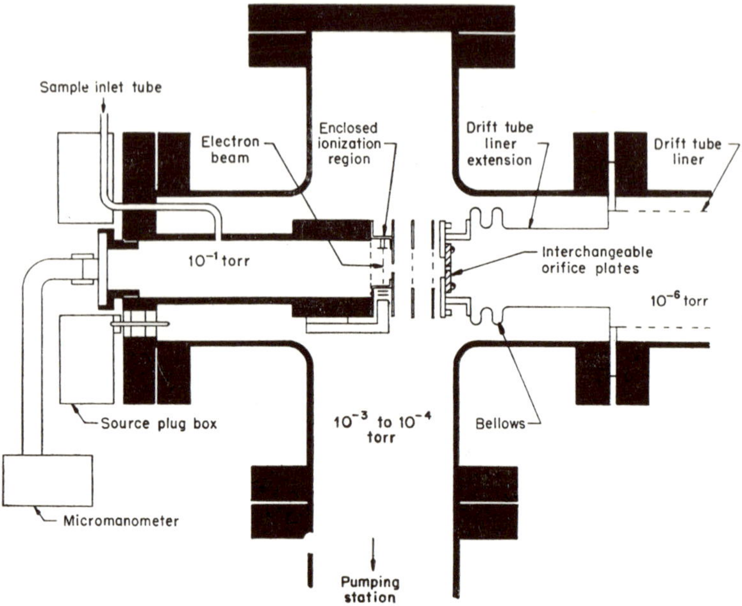

FIG. 13. Ion molecular source.

type of measurement, thus eliminating one of the most elusive variables in ion–molecule reaction studies. An oversized pump is utilized to handle the tremendous amount of gas pouring out of the high pressure region. The people at Wright Patterson Air Force Base (J .Futrell *et al.*) have been studying ion–molecule reaction studies at about 1 torr pressure and getting results not previously discovered. The addition of a supplementary heater for the source was a most important step for doing ion–molecule reactions. Keeping the source region warm minimizes coating and insulation-type build up on the inside of the source which often caused measurements to go awry.

The simultaneous multiple channel arrangement allows measurements on very small thermally ionized samples with better accuracy than instruments which monitor only a single species at a time (Fig. 14). The thermal ionization inlet system is a variation of the familiar direct inlet probe.

Figure 15 shows two different types of ion sources; the standard source (where the sample gas is introduced through a gas

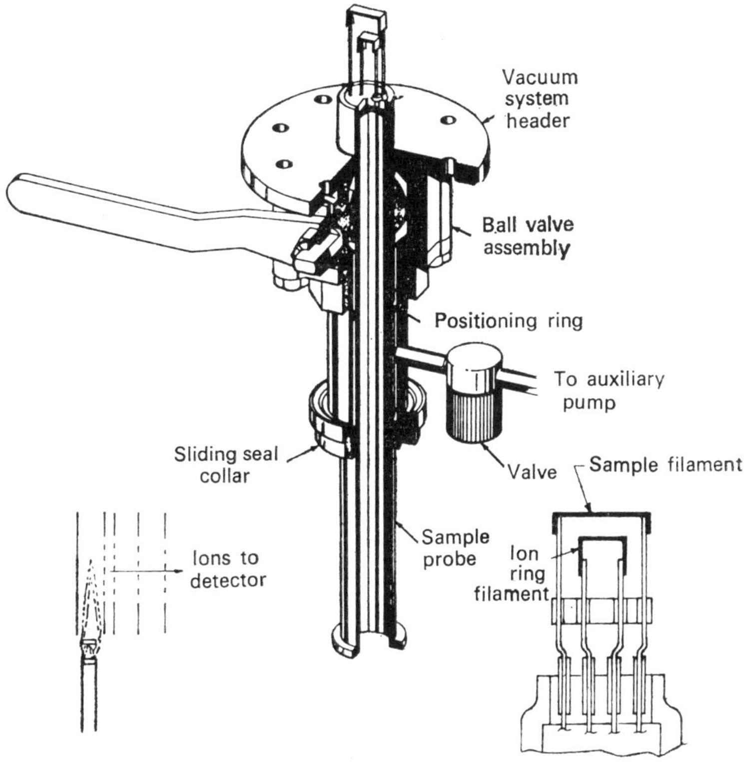

FIG. 14. Thermal (surface) ionization probe).

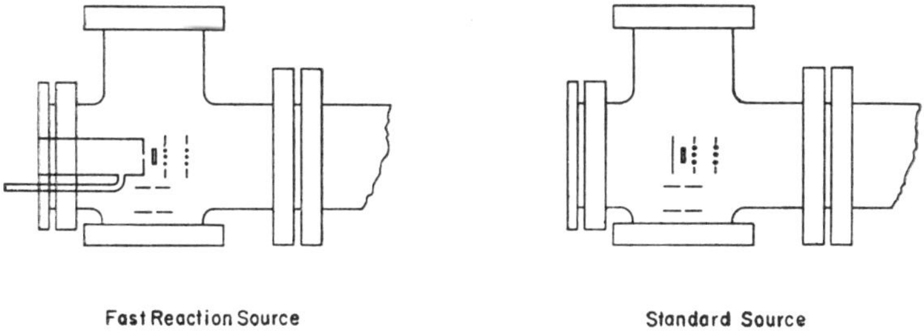

FIG. 15. Ion sources.

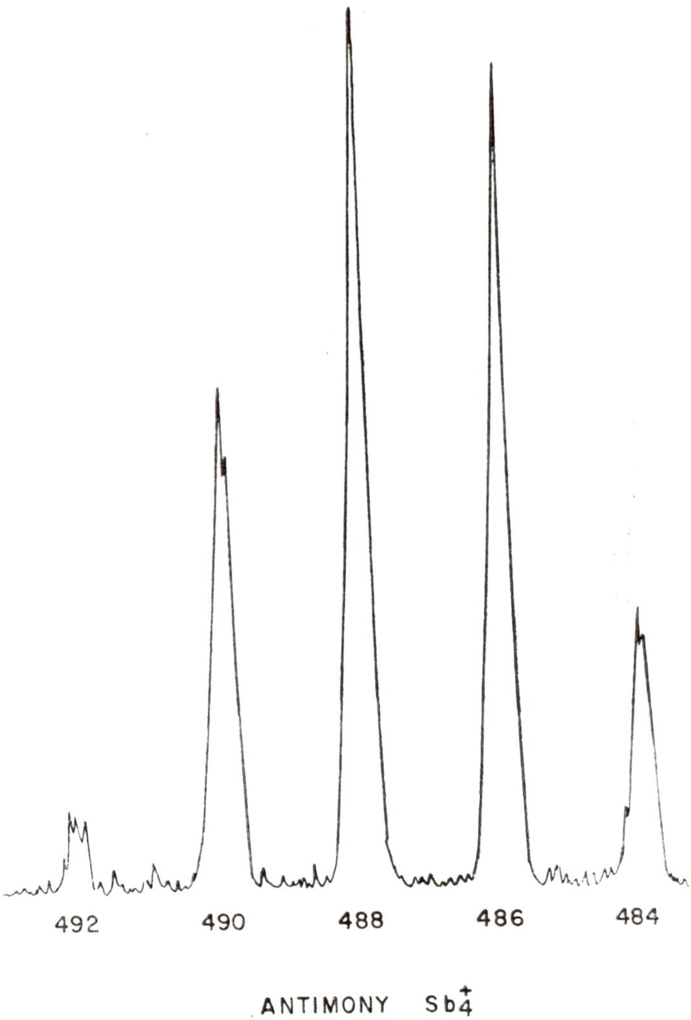

FIG. 19. Spectrum showing resolution on Model 3012 TOFMS.

FIG. 16. TOFMS nude source.

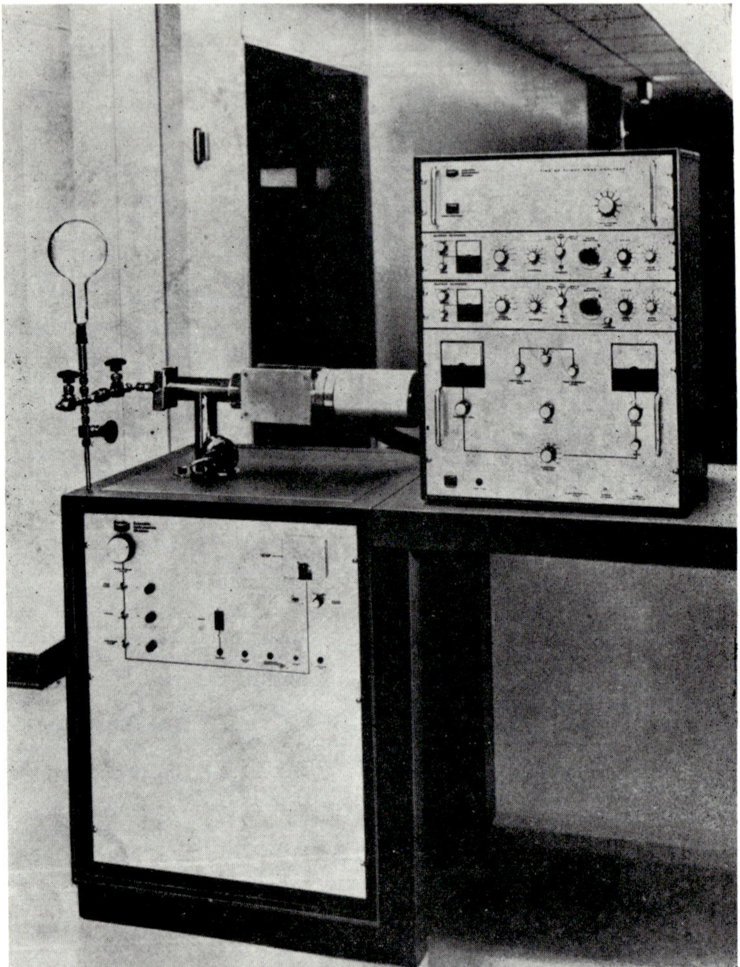

Fig. 17.

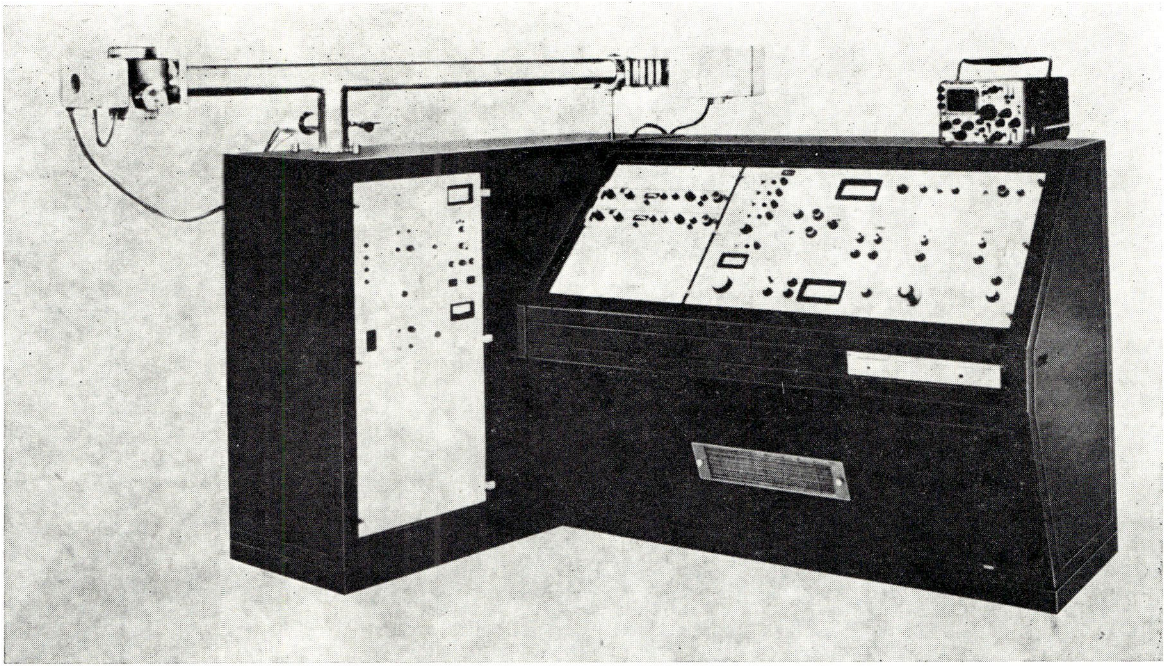

Fig. 18. Model 3015 TOFMS.

inlet, a direct inlet, Knudsen cell, etc.) and the fast reaction source designed by Kistiakowsky. The latter source is about 3 cm inside diameter, therefore the outside diameter of the shock tube, flash lamp, or other apparatus for getting the sample close to the ionizing electron beam, is somewhat restricted. Dr. Bauer of Cornell was first to point out the advantages that could be realized with the arrangement shown in Fig. 16. This is called the nude source and is an ion source which projects outside the vacuum system. The source is enclosed in an appropriate vacuum envelope and the reaction chamber is so arranged that the sample can be injected from the end, as with a fast reaction source, or orthogonal to the electron beam and the ion flight axis. This configuration is especially well suited for shock tube work and fast reaction studies.

Recently, a physicist at Bendix, Charles Moorman, discovered that the flight tube length factor canceled out of the basic TOF equation, and consequently the energy focusing of the TOFMS does not depend primarily on the length of the drift tube. So a new TOFMS was born—the Bendix Model MA-1 (Fig. 17). This instrument features a 20 mm (8 in.) flight path and still achieves a resolution in excess of 200. The developments of the 100 Mc/s oscilloscopes, such as the Tektronix 453A, and avalanche transistor pulse circuits, make possible the scanning of closely spaced peaks from this mass spectrometer. The fact that the peaks come out of this instrument about 5 nanosec apart gives sufficient time resolution to differentiate between mass-to-charge species at 200.

It was found that the energy level of the instrument would have to be about 4000 V to achieve resolution on energetic molecules of mass 1000. Also, resolution can be achieved on a broad spectrum of organic species to mass 600–700 with about 3000 V energy. However, there is a definite relationship between the accelerating and drift voltages and the maximum resolution obtainable over a broad spectrum of organic molecules. The fact that the MA-1 has a 1000 V ion energy limits the instrument to a resolution on organics of about 250. To date, without using time-lag focusing or other special techniques, Bendix has achieved a resolution of about 150 on organic species and just over 200 on inorganics with this instrument. The mass range is about 1000. The analyzer includes the ion source, the drift tube, and the magnetic electron multiplier and is all mounted on a single vacuum header. It is about 16 in. overall and will operate to at least 250°C with little loss of resolution.

Figure 18 shows the Bendix Model 3015, which is the most versatile TOFMS One of its outstanding features is that analog recordings of both positive and negative ions are possible. One can be analyzing the positive ion spectrum of a sample, shut off the high voltage, flip the polarity switch, turn the high voltage back on, and with little additional adjustment, have a negative ion spectrum. The Bendix Model 3012 is a positive ion only version of the 3015.

Both the Model 3012 and 3015 were designed to give the ultimate performance possible for a TOFMS. With the continuing

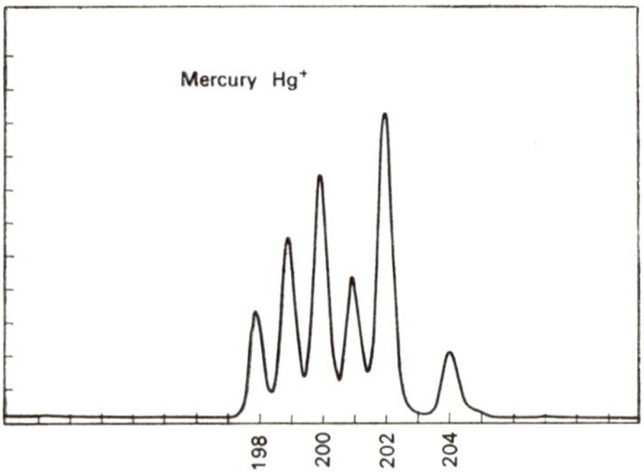

FIG. 20.

improvement in resolution, stability and sensitivity shown to date, Bendix feels that these instruments have already surpassed every design goal originally set for them. Figure 19 shows antimony (Sb_4^+) spectrum with a resolution of over 1000. Of course, this is with an inorganic compound. With organic materials the 3012 and 3015 achieve 500 to 700 resolution by the $M/\Delta M$ at 10% peak height definition. Figure 20 shows a mercury scan from the MA-1 for comparison.

Further improvements in resolution and sensitivity of all versions of the TOF mass spectrometer will be achieved. These will be important for all types of analyses. The improved resolution, plus the recent breakthrough in timing pulse stability, will permit computer data reduction of spectra in the near future. For routine analysis, the operator can watch the panorama on the oscilloscope, while the spectra are digested by the computer. But we need not worry. The operator will still be required to load the syringe and inject the sample.

APPLICATIONS OF TIME-OF-FLIGHT MASS SPECTROMETERS

FLASH PHOTOLYSIS REACTIONS

R. T. MEYER†
Sandia Laboratories, Albuquerque, New Mexico, U.S.A.
and D. PRICE
Department of Chemistry, The University of Salford

THE initiation of chemical reactions by the application of high energy radiation from flash tubes is well known. The technique is capable of producing high transient concentrations of atoms or free radicals and of initiating fast reactions. The basic technique as pioneered by Norrish and Porter[(1)] used the experimental arrangements shown in Fig. 1. The photolysis lamp gives a rapid intense light flash when the associated bank of condensers is discharged.[(2)] The energy output, in joules (J), is given by $\frac{1}{2}CV^2$, where C is the condenser capacity in microfarads and V is the condenser potential in kilovolts. Normally, the flash energies range from 500 J upwards whilst the flash lifetimes vary from 1 to 200 μsec depending upon the circuit characteristics.[(1)]

The energy from the photolysis flash is reflected into the relevant chemical system in the reaction vessel. It is essential that the chemical system absorbs the flash energy. Sensitizers may or may not have to be used to accomplish this. The intensity of the photolysis flash is such that unless precautions are taken, e.g. the heat capacity of the system is made large, the light energy absorbed by the system will be rapidly converted to heat either by chemical reaction or collision relaxation. If no precautions are taken, then it is possible to produce, virtually instantaneously, temperatures of 2000–3000°C within the reacting system. Chemical reactions so produced are pyrolytic rather than photochemical in nature and may be referred to as flash pyrolysis reactions. By allowing the temperature to rise in a potentially reactive system, both pyrolytic and explosive processes can be studied.

† This work was supported in part by the United States Atomic Energy Commission.

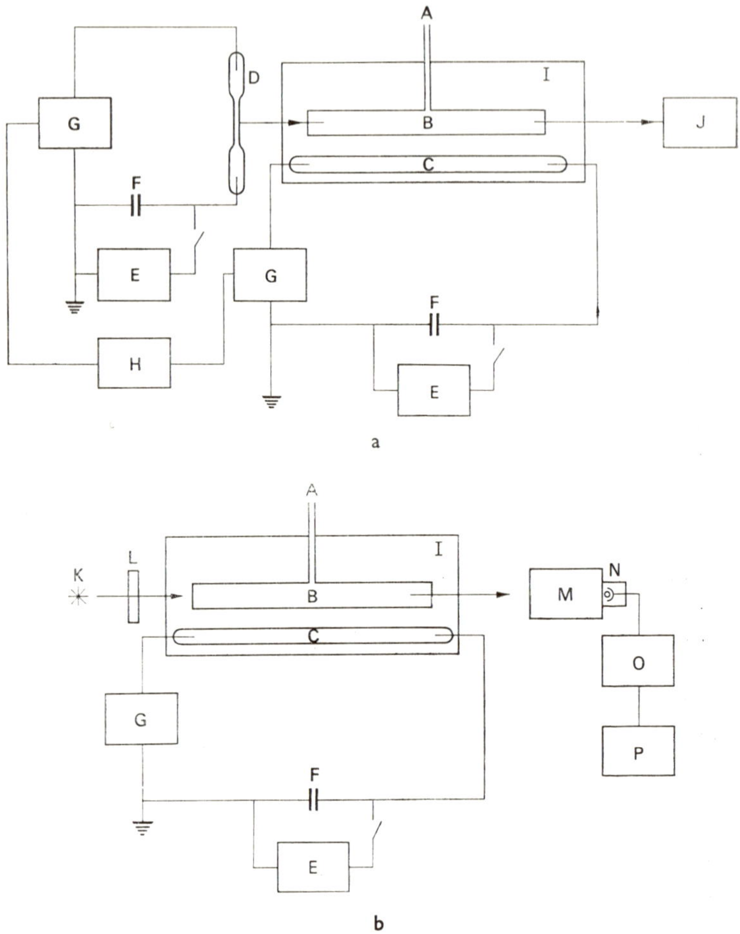

FIG. 1. (a) Spectroscopic flash photolysis. (b) Kinetic flash photolysis. A, reactant inlet; B, reaction vessel; C, photolysis lamp; D, spectroscopic lamp; E, charging unit; F, condensers; G, firing unit; H, delay unit; I, reflector; J, spectrograph; K, monitoring light source; L, filter; M, monochromator; N, photomultiplier; O, amplifier; P, oscilloscope and camera.

Solid, liquid and gaseous systems have been investigated by this technique. If the heat capacity of the reaction system is increased, in the gas phase by the addition of some hundred-fold excess of inert diluent gas or by the solvent in the case of solutions, then the temperature rise can be kept below 10°C and a wide range of photochemical reactions involving atoms and free radicals can be studied. This is the flash photolysis technique.

A flash photolysis reaction, once initiated, is followed by monitoring the change in concentration with time of the appropriate species present. In many ways absorption spectroscopy has proved ideal for this purpose. Two techniques have been developed. In one, the spectroscopic flash lamp shown in Fig. 1a is fired a short time interval after the initiation of reaction by the photolysis lamp. Thus, an absorption spectrum of the reacting system is recorded photographically in the spectrograph. This process is repeated for various reaction time intervals so that a series of absorption spectra versus time are obtained.[(1)] In the second technique, Fig. 1b, a continuous light source, in conjunction with a monochromator and photomultiplier, continuously monitors the reacting system at any desired characteristic wavelength. Absorption spectroscopy is a satisfactory method of following flash photolysed reactions in that, while it does not disturb the system, it is able in many cases to provide a qualitative and a quantitative estimate of the species in the reacting system. The application of absorption spectroscopy is limited in that many important reaction intermediates either do not absorb strongly enough for their detection in the accessible region of the spectrum or have unidentifiable continuous spectra. Successful monitoring of flash photolysed reactions by mass spectrometry, with the possibility of qualitative identification of reaction species and quantitative analysis of complex mixtures, would be advantageous in remedying some of the deficiencies in the absorption spectroscopy technique. The use of a magnetic deflection instrument would involve use of photographic plate recording or monitoring a single mass peak, due to the species of interest, at an electron multiplier detector. The latter technique has been used by Strausz to study reactions of selenium atoms produced by the flash photolysis of carbonyl selenide[(3)] and also the photolysis of tricarbonylcyclobutadienyliron.[(4)] This system has the disadvantage that only one peak in the mass spectrum can be recorded at any one time. The application of the time-of-flight mass spectrometer, with its ability to produce a mass spectrum in 10^{-4} sec or less, is obviously desirable for monitoring flash photolysed reactions.

Kistiakowsky and Kydd[(5)] were the first to apply time resolved mass spectrometry to flash photolysed gas reactions. A simple sketch of their apparatus is shown in Fig. 2. The reaction system was continuously sampled via the pinhole in the gold foil separating the reaction cell from the ion source. Reaction pressures were of the order of 1 torr with the pressure in the ion source being 10^{-5} torr. The mass spectrometer had a pulse rate of 20 kHz,

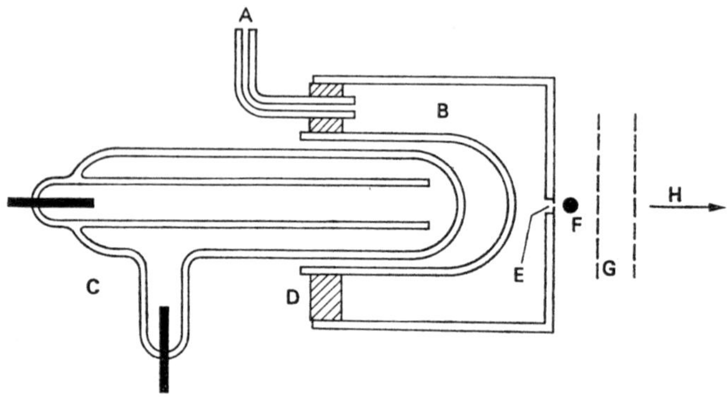

FIG. 2. Flash photolysis cell–ion source combination as used by Kistiakowsky and Kydd.[5] A, reactant inlet; B, reaction vessel; C, flash lamp; D, teflon plug; E, pinhole; F, electron beam; G, grids; H, flight tube.

so that the mass spectrum of the reacting system was displayed on the monitoring scope every 50 μsec. The spectra were photographed on a drum camera, the camera shutter and flash tube trigger being synchronized to record the mass spectra produced immediately after the flash. Inert gases were used to reduce the thermal effects of the absorbed flash energy. The neon *m/e* 20 peak was also used as reference to allow for variation in the mass spectrometer sensitivity. Two studies made with this apparatus were reported.

PHOTOLYSIS OF KETENE

Ketene was flash photolysed in the presence of neon and, in one experiment, sulphur hexafluoride. The mass spectra showed that ethylene and carbon monoxide were present within 50 μsec of the flash being fired. No strong peak was observed at *m/e* 14, indicating that the methylene radical was not one of the main species present. However, the technique was not good enough for this to be proved conclusively. The reaction mechanism proposed to account for these experimental observations was:

$$CH_2CO + h\nu \longrightarrow CO + CH_2 \quad (a)$$

$$CH_2 + CH_2CO \longrightarrow C_2H_4 + CO \quad (b)$$

The results indicate that reaction (b) is very fast with a reaction probability on collision of at least 10^{-2}. A peak at m/e 55 was observed to appear immediately after the flash. This disappeared from the reaction cell after several hundred microseconds. It was suggested that this was due to a radical or molecule of empirical formula C_3H_4O present as an initial product of the reaction under these conditions.

PHOTOLYSIS OF NITROGEN DIOXIDE

The flash photolysed decomposition of nitrogen dioxide in the presence of neon was also reported. Approximately 5% decomposition of the nitrogen dioxide occurred. No significant change in oxygen atom concentration was observed after the flash while the oxygen molecule concentration reached a maximum within 100 μsec. The production of oxygen atoms is known to be a primary step in the photochemical decomposition of nitrogen dioxide. The failure to observe a change in oxygen atom concentration was ascribed to poor light absorption by the system resulting in the change in oxygen atom concentration being below the detectable limits. The rapid increase in oxygen molecule concentration indicated that the reaction of oxygen atoms with residual nitrogen dioxide to form oxygen molecules and nitric oxide[6]

$$O + NO_2 \longrightarrow O_2 + NO$$

is very fast with the lower limit of reaction probability on collision being about 10^{-2}. No nitrogen trioxide was detected. Significant amounts of nitrous oxide were formed at the same rate as the oxygen molecules. It was suggested that the primary dissociation of nitrogen dioxide by radiation of shorter wave lengths results in the formation of "hot" molecules of nitric oxide, which then react according to

$$NO^* + NO_2 \longrightarrow N_2O + O_2$$

The value of the exploratory studies by Kistiakowsky and Kydd is that they demonstrated the feasibility of monitoring fast reactions by time-of-flight mass spectrometry. Their spectrometer was a prototype and suffered two principal disadvantages: (1) low mass resolution ($\sim$50), and (2) low sensitivity (1 part in 700). Another major limitation was associated with their flash photolysis apparatus. The intensity of the flash lamp was insufficient to cause greater than 10% photodecomposition of the reactant gases at the low pressures used. As a result, the high concentrations of atoms and radicals typical of the works of Porter[1a] were not achieved. The concentrations that were produced were too low for detection with the limited sensitivity of the spectrometer at that time. These restrictive factors led many prospective users to abandon the idea of coupling flash photolysis and time-of-flight mass spectrometry for studies of fast gas phase reactions.

In late 1960, at Sandia Laboratories, an effort was initiated to design a flash lamp and reaction apparatus comparable in efficiency with the systems used in flash absorption spectroscopy and to improve the sensitivity of the time-of-flight mass spectrometer.[7] By that time, the mass resolution of the Bendix time-of-flight mass spectrometer had been substantially improved and low resolution was no longer a serious problem. Several geometries[7–13] of a combined flash lamp, reaction vessel and spectrometer ion source were explored in an effort to attain efficient coupling of the ultraviolet radiation to the reaction zone and close proximity of the reaction zone to the electron beam. The finger and thimble configuration shown in Fig. 3 proved the most successful design. The flash lamp utilized a coaxial discharge path between two quartz tubes. A Suprasil-grade quartz window was sealed onto the end of the outer tube. This seal was the weak point structurally and limited the maximum current density that could be discharged. The lamp was filled with 5 torr of krypton or xenon, which was an optimum pressure for short duration pulses. Input energies and flash durations ranged from 2570 J in 260 μsec to 908 J in 12 μsec.[8–10]

The two-piece design of the flash lamp and reaction vessel offered several advantages. Flash lamps with various transmission windows and with various gas fillings could be used interchangeably with reaction vessels having different size orifices and different glass structures. The early versions of the vessel were con-

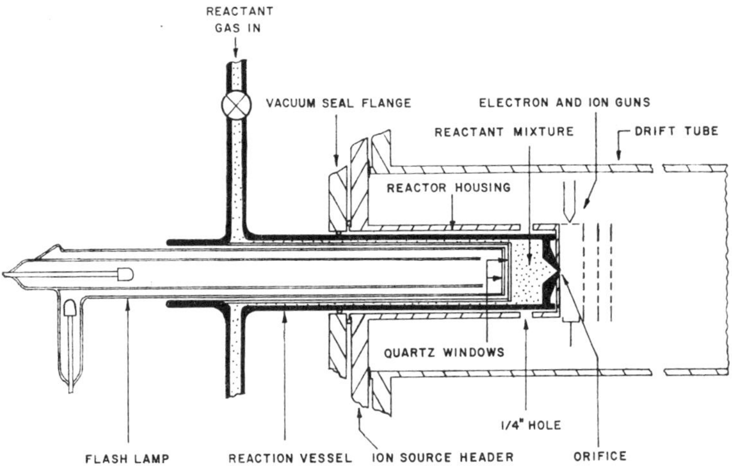

FIG. 3. Geometry of coaxial flash lamp and opaque black-glass reaction vessel adapted to ion source of Bendix time-of-flight mass spectrometer.

structed of quartz and pyrex, to minimize surface reactions,[8–13] and were surrounded with gold foil reflectors. Later, a vessel with an opaque black glass outer shell was designed to eliminate the flash desorption of contaminates from the stainless steel wall of the ion source reactor housing.[14] The addition of the outer tube extension (to the left of the inlet gas arms in Fig. 3) allowed the use of a gas or liquid light filter when a tube coupling was applied between the vessel and lamp.

The effective reaction zone was the 1·5 cm long by 2·9 cm diameter cylindrical volume between the Suprasil window and the sampling orifice. The orifice diameters ranged from 10 to 100 μm. With the smallest orifice and with an auxiliary pumping station on the ion source cross of the Bendix model 14–101 spectrometer, reaction vessel pressures as high as 100 torr were obtained while the spectrometer pressure was maintained at 10^{-5} torr.[10, 11]

This improvement in flash lamp design and reaction vessel configuration was demonstrated by nitrogen dioxide actinometry.[10, 11] For a single 2570 J flash, 80% decomposition of nitrogen oxide was obtained. This was correlated with an incident ultraviolet light flux of $1{\cdot}0 \times 10^{18}$ quanta cm^{-2}. Fluxes two or three times larger were also recorded. Therefore, the apparatus should be capable of producing 5% decomposition of a substance with an exponential extinction coefficient of only 40 l. mole^{-1} cm^{-1}.

Several factors contributed to a vastly improved spectrometer sensitivity.[9–11] All related to increasing the number of ions that are produced, transmitted and collected per cycle of the time resolved analysis. (1) The aforementioned reaction vessel design located the sampling orifice 2 mm from the centre of the electron beam. This produced a larger number of ions since the electron beam intersected the gas flow in a region of high molecular density. (2) The number of interacting electrons was increased by lengthening the electron beam pulse duration from 0·25 to 1·4 μsec. (3) The nickel mesh of the three ion grids was replaced with tightly stretched and welded molybdenum mesh to reduce the dispersion of ions during draw-out and acceleration.[15] (4) A variable, intermediate accelerating potential was applied to the second ion grid. The absolute value ranged from -300 V to -1500 V and depended on the overall ion source conditions, gas sample and mass range. (5) The several other pulse shaping and ion focusing parameters were adjusted to maximize the ion intensities for each set of experiments and for the mass range of interest.[11]

From measurements of argon ion statistics and assuming a Poisson distribution of molecules, it was shown that approximately 100,000 ions/cycle were produced, transmitted and collected with the improved spectrometer (operated at 20 kHz pulse rate, 70 eV electron energy and 25 μA trap current).[10, 11] This provided a relative sensitivity of 1 part in 100,000 and an absolute sensitivity for a reactive species of 5×10^{-8} mole l.$^{-1}$, when a total sample pressure of 10 torr was employed. Bimolecular reactions having rate constants of 10^8 to 10^{10} l. mole^{-1} sec^{-1} are easily resolved. Under favourable conditions it should be possible to observe reaction rates approaching 10^{11} l. mole^{-1} sec^{-1}.

An improved oscilloscope display and revolving drum camera technique was utilized for recording the time resolved mass spectral data.[16] Data reduction of the film records was facilitated by a computer programme, which made corrections for systematic instrumental variation, impurity contribution and dissociative ionization and provided a final output of ion intensity versus time for each component in the chemical reaction.[17]

Chemical systems which have been investigated with this improved instrumentation include the following: the nitrogen dioxide sensitised reaction between hydrogen and oxygen;[9, 12, 13] the decomposition of nitromethane;[18] the termolecular recombination of iodine atoms in the presence of nitric oxide;[19] and the reaction of excited iodine ($^2P_{1/2}$) atoms with methyl iodide.[19, 20]

NITROGEN DIOXIDE SENSITIZED REACTION BETWEEN HYDROGEN AND OXYGEN

The most significant result of this study was the detection of the hydroxyl radical as a transient, reactive species. It was observed simultaneously with the formation of water as a final product. Reaction initiation occurred by photodecomposition of the nitrogen dioxide into nitric oxide and oxygen atoms and by the conversion of excess light energy into thermal energy. The kinetic data indicate that the subsequent steps were in agreement with the following mechanism.[6, 21, 22]

$$NO_2 + h\nu \longrightarrow NO + O + \text{energy} \quad \text{(a)}$$

$$O + NO_2 \longrightarrow NO + O_2 \quad \text{(b)}$$

$$O + H_2 \rightleftharpoons OH + H \quad \text{(c)}$$

$$H + O_2 \rightleftharpoons OH + O \quad \text{(d)}$$

$$H + NO_2 \longrightarrow OH + NO \quad \text{(e)}$$

$$OH + H_2 \longrightarrow H_2O + H \quad \text{(f)}$$

$$OH + OH \longrightarrow H_2O + O \quad \text{(g)}$$

Supporting experimental evidence is contained in Table 1 and Fig. 4 and 5. For the pure nitrogen dioxide system, reactions (a) and (b) are indicated by the measured nitric oxide : molecular oxygen final ion intensity ratio of 2·5. A final product concentration ratio of 2·0 has been demonstrated by Ford[23] for the photolysis at low pressure, where termolecular reactions are insignificant. The detection of the OH radical and the measured correlation between its rate of decay and the rate of formation of water, shown in Fig. 4, suggest the occurrence of reactions (c), (f) and (g). The presence of H atoms is indicated by the increase in the amount of nitric oxide produced when hydrogen is added

TABLE 1. *Changes in reactant and product concentrations for various reactant mixtures*

Product formation data are parent ion intensity changes, expressed per unit amplitude of the initial ion intensity of nitrogen dioxide; ionization efficiency correction factors have not been applied.

Reactant mixture	Total pressure (torr)	NO_2 decomposed (mole fraction)	Product formation		
			NO	O_2	H_2O
NO_2	6·3	0·79	2·5	1·0	0
NO_2+H_2	5·3	0·78	3·3		1·9
$NO_2+3\,H_2$	10·0	0·74		0·24	1·7
$NO_2+3\,H_2+O_2$	12·6	0·69	2·9	−0·37	3·8
$NO_2+2\,H_2+2\,O_2$	13·8	0·62	7·7		3·9
$NO_2+3\,H_2+4\,O_2$	18·7	0·61	2·6		4·1

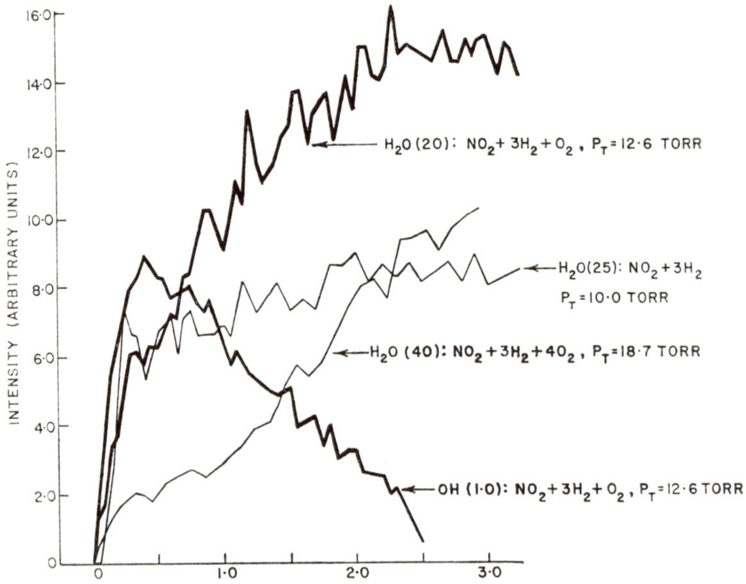

FIG. 4. OH radical and H_2O ion intensities versus time from the flash photolysis of NO_2, H_2 and O_2 at various pressures; the number in each parenthesis, e.g. (1·0), represents the relative ion intensity scale factor in units/division. Flash photolysis starts at time zero and has a $1/e$ duration of 0·25 msec. Initial rate of formation of H_2O decreases as the O_2 to H_2 ratio and total pressure P_T increase.

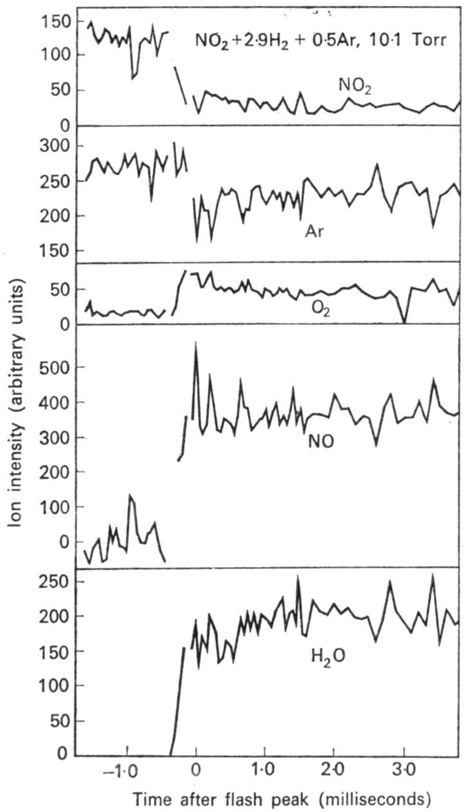

FIG. 5. Time resolved record of decay of NO_2 and of formation of O_2, NO and H_2O as final products. O_2 data show initial rapid rise from NO_2 decomposition, followed by a decrease due to reaction with hydrogen atoms. All the data points have been referenced to the argon data to correct for the systematic density decrease due to flash heating and thermal expansion.

to nitrogen dioxide and the chain propagating reactions (c), (f) and (e) occur. In general, the formation of nitric oxide is rapid, following the decay of nitrogen dioxide very closely. The depletion of molecular oxygen is slower and parallels the formation of water. In fact, when molecular oxygen is not initially present, its rapid formation from nitrogen dioxide is observed prior to its direct participation as a reactant in reaction (d) (Fig. 5). Other investigators have shown that reaction (d) is endothermic.[24]

This is confirmed by the experiments with larger oxygen : hydrogen ratios, for which the excess oxygen molecules act as a thermal sink. The rate of water formation is seen in Fig. 4 to decrease with increasing oxygen content. Furthermore, the measured consumption of oxygen molecules and the twofold increase in the total water produced, when oxygen is added to a nitrogen dioxide plus hydrogen mixture (see Table 1), conclusively demonstrate reaction (d).

The overall temperature reduction produced by adding oxygen and increasing the total pressure was correlated with increases in the induction period for hydroxyl radical formation. In addition it was ascertained that an increase in flash intensity leads to reaction acceleration by increasing the initial supply of oxygen atoms.[(13)] The latter was determined by comparing the rates of hydroxyl radical and water formation for reactant mixtures for which the gas temperatures were approximately equal.

DECOMPOSITION OF NITROMETHANE

A cursory examination of the flash photolysis of nitromethane was performed because of its similarity to the nitrogen dioxide–hydrogen–oxygen system. The photodecomposition of nitromethane is of general interest, however, since the primary mode of dissociation has been uncertain for some time. Four possible dissociation steps are:

(a) Cleavage of the C–N bond leading to the generation of a methyl radical and nitrogen dioxide;[(25)]

$$CH_3NO_2 + h\nu \longrightarrow CH_3 + NO_2$$

(b) Dissociation of a N–O bond, producing oxygen atoms:

$$CH_3NO_2 + h\nu \longrightarrow CH_3NO + O$$

(c) Intramolecular rearrangement to the nitrite isomer,[(26)] followed by dissociation to the methoxyl radical:

$$CH_3NO_2 + h\nu \longrightarrow CH_3ONO \longrightarrow CH_3O + NO$$

(d) intramolecular rearrangement and decomposition to formaldehyde:[(27)]

$$CH_3NO_2 + h\nu \longrightarrow CH_3ONO \longrightarrow CH_2O + NOH$$

Figure 6 shows data obtained in a flash photolysis experiment. This experiment was performed with a total pressure of 1 torr under adiabatic conditions. It is seen that the amplitude of the parent mass peak for nitromethane (mass 61) decreases rapidly after the flash, then rises slightly and decreases again. The interpretation of this behaviour is still unresolved. Masses 28 and

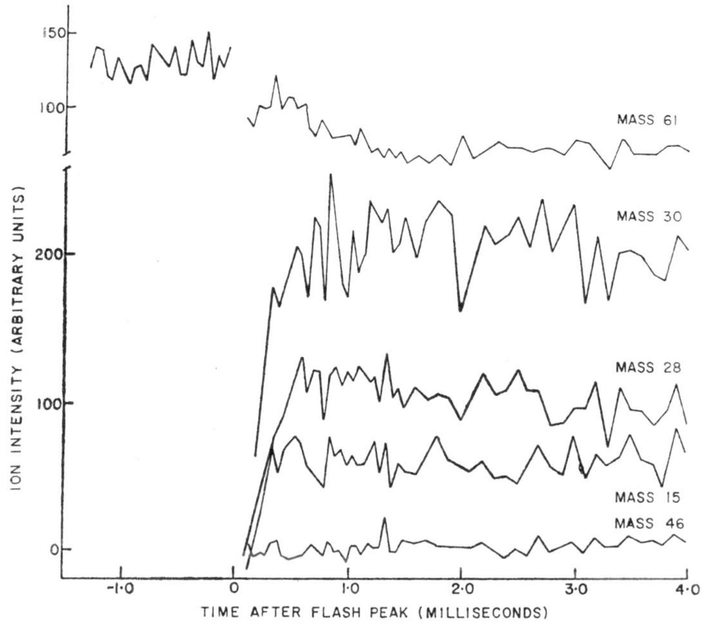

FIG. 6. Flash photolysis of nitromethane; $CH_3NO_2 + 0{\cdot}5$ Ar, 1 torr. Time resolved records show decay of nitromethane (mass 61) and formation of carbon monoxide (mass 28), nitric oxide (mass 30) and mass 15. Nitrogen dioxide (mass 46) is absent as a transient.

30 both increase with time, which probably correspond to the formation of carbon monoxide and nitric oxide, respectively, as final products. It is not clear at this point what final products contribute to the significant increase in mass 15. Most interesting is the absence of nitrogen dioxide (mass 46) as a final product or even as a transient very early in the reaction. At least three explanations are possible: (1) the primary act does not involve splitting of the C–N bond; (2) if the C–N bond does split, the nitrogen dioxide immediately disappears by thermal or photo-

chemical decomposition; and (3) the nitrogen dioxide forms and subsequently reacts in a time interval shorter than the earliest observation time.

According to recent studies by Napier and Norrish,[(28)] it appears that explanation (3), a rapid reaction of nitrogen dioxide, is most reasonable. These authors showed that the flash photolysis of nitromethane results in the formation of methyl radicals and nitrogen dioxide. The fragments then recombine to form methyl nitrite and also disproportionate to yield methoxyl radicals and nitric oxide. Under adiabatic conditions they did not observe nitrogen dioxide, but did detect small radicals such as NH and CN and also nitroxyl. They suggested that nitrogen dioxide is an oxidizing agent for various intermediate compounds such as formaldehyde or methanol.

TERMOLECULAR RECOMBINATION OF IODINE ATOMS IN THE PRESENCE OF NITRIC OXIDE

The study of atomic iodine recombination was undertaken as a test of the flash photolysis and time resolved mass spectrometry apparatus to generate quantitatively reliable data. The source of iodine atoms was the photolysis of methyl iodide. Nitric oxide was chosen as the third body because it is a very effective quencher of excited iodine atoms[(29)] and is the most efficient chaperon for termolecular recombination.[(30)] The former need arises since an excited iodine ($^2P_{1/2}$) atom is produced in the photodecomposition of methyl iodide, and the latter need arises because of the possibility of competing iodine atom reactions.

Experiments were performed using a phosphoric acid-wetted, opaque reaction vessel. The nitric oxide/methyl iodide ratio and the total pressure were varied. Figure 7 shows the second-order plots for two experiments which indicate that the reaction rate does indeed increase upon the addition of more nitric oxide. The average value of the rate constant from eleven experiments is $2{\cdot}4 \pm 1{\cdot}4 \times 10^{13}$ l^2 mole^{-2} sec^{-1}. The value reported by Porter *et al.*[(30)] is $3{\cdot}0 \times 10^{13}$ l.^{2} mole^{-2} sec^{-1}. This good agreement establishes that the termolecular recombination of iodine atoms has been identified and measured by the technique of time resolved mass spectrometry.

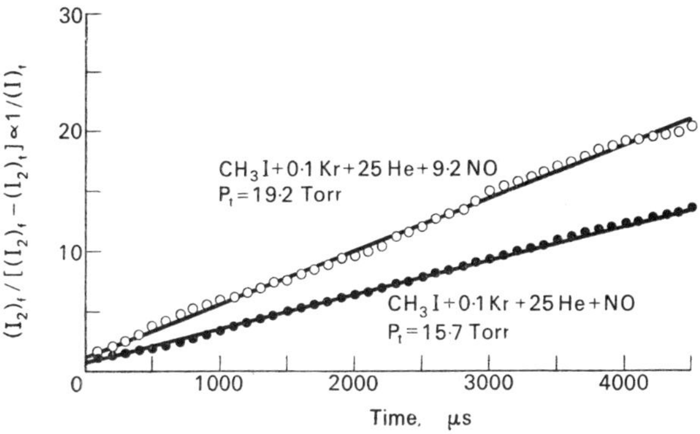

FIG. 7. Second-order plot of I_2 formation due to termolecular atom recombination in presence of NO. Flash parameters: 908 Joules, 12 μsec duration. Phosphoric acid-wetted, opaque reaction vessel.

REACTION OF EXCITED IODINE ($^2P_{1/2}$) ATOMS WITH METHYL IODIDE

Previous studies have fairly well established that the photodissociation of methyl iodide produces a methyl radical and a metastable excited iodine ($^2P_{1/2}$) atom.[31–33] The methyl radical carries off excess translational or vibrational energy and undergoes hot radical reactions. The fate of the excited iodine atom has usually been described as collisional deactivation and termolecular recombination to molecular iodine. From time resolved mass spectrometric studies, evidence has been obtained for the chemical reactivity of excited iodine atoms.

Figure 8 is a plot of molecular iodine ion intensity as a function of time for various gas mixtures.[20] Data points are plotted from 75 μsec to 4 msec after the flash peak, with a final data point at 5 sec. It has been found that from 20% to 75% of the total final amount of molecular iodine is formed within 100 μsec after the flash. For the concentration parameters of these experiments, this rapid formation is not explicable in terms of the well-known termolecular recombination. Nor is it explicable in terms of any elevated-temperature reaction.

More recent studies have consisted of measuring the rate of molecular iodine formation as a function of the ratio of methyl iodide diluted with helium and of the total pressure.[14] This

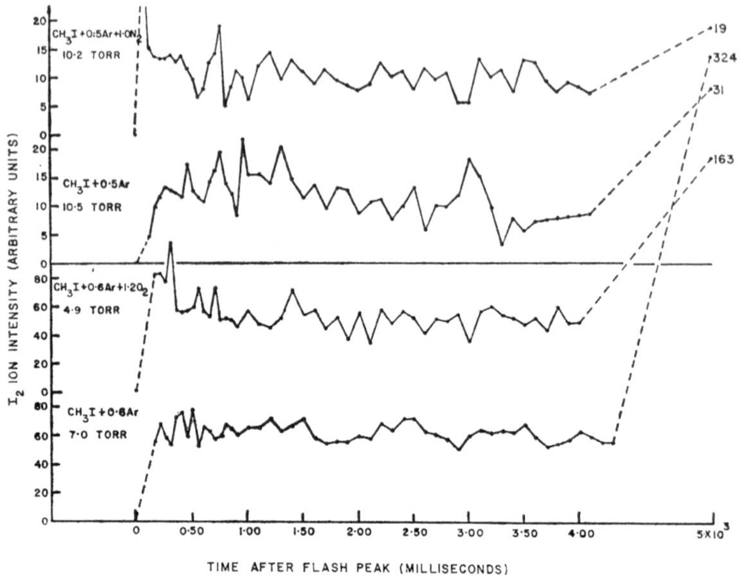

FIG. 8. Flash photolysis of methyl iodide; time history of molecular iodine formation in the presence and absence of N_2 and O_2. The upper pair of plots was obtained using a flash duration of 18 μsec; first data points are recorded at 75 μsec after the flash peak. A flash duration of 90 μsec applies to the lower pair, with the first data points recorded at 170 μsec. Data points at 5 sec were obtained from a 1/50 sec Polaroid film exposure; they represent the average final ion intensity of I_2.

was done to observe any temperature dependence of the initial rate. Helium was chosen as the diluent to reduce collisional deactivation of the iodine ($^2P_{1/2}$) atoms. A rapid initial reaction was again observed, which revealed a weak temperature dependence ($E_{act} \simeq 2$ kcal mole^{-1}) over the temperature range 335–605°K. For conditions under which the total methyl iodide decomposition was 10%, the initial data agreed with a first order kinetic behaviour for iodine atoms.

From these results, it has been concluded that the following excited iodine atom reaction occurs in the flash photolysis of methyl iodide:

$$I(^2P_{1/2})+CH_3I \longrightarrow I_2+CH_3+3\text{ kcal}$$

Pimentel and co-workers[33–36] have suggested an analogous reaction to explain the quenching of laser emission ($^2P_{1/2} \rightarrow {}^2P_{3/2}$)

at elevated temperatures in the flash photolysis of trifluoroiodomethane. Such a reaction provides for quenching in three ways: (1) removal of the emitter, iodine ($^2P_{1/2}$) atom; (2) depletion of the source of the emitter, methyl iodide or trifluoroiodomethane; and (3) production of the effective quencher, molecular iodine.

The studies on methyl iodide decomposition have also identified the surface recombination of iodine atoms under certain conditions. The proper conditions were unwetted walls of the reaction vessel, moderately low pressures and a gas temperature near 300°K. No rapid initial formation of iodine molecules was observed and the slower rate of molecular iodine formation was quantitatively correlated with a diffusion-controlled, surface recombination. This result suggests the applicability of the reaction vessel geometry and the time resolved mass spectrometry technique to other surface catalysed reactions.

Flash heating has been a significant factor in the studies mentioned. While it was desired in the nitrogen dioxide–hydrogen–oxygen system to facilitate the "chain-thermal" reactions, it was avoided in the study of the termolecular recombination of iodine atoms in the presence of nitric oxide to achieve an isothermal reaction rate. For the iodine ($^2P_{1/2}$) atom–methyl iodide reaction, flash heating was utilized to measure the temperature dependence of a rate process. In all of these systems, which employed an open-ended reaction vessel, argon or an equivalent inert gas was present in small amount as a qualitative sensor of gas temperature. It was observed that argon density was inversely proportional to the gas temperature in the reaction zone (for an example, see Fig. 5). Although inert gas density has not been used to obtain quantitative measurements of temperature, the gross observations so far are in general qualitative agreement with heat capacity estimates of reaction temperature.[11–14]

The use of a large excess of inert gas to achieve isothermal conditions is restricted by the spectrometer sensitivity. Since only a limited number of ions can be produced and detected ($\sim$100,000 per cycle) by the spectrometer, the use of an inert gas reduces the number of ions of all other reactant, product and intermediate species that can be detected. As an example, if an inert gas to reactant gas ratio of 100 to 1 is used, the effective sensitivity is reduced to 1 part in 1,000. Helium is often a good choice for an inert gas, because it has a low ionization efficiency, is readily pumped with a diffusion pump and is easily defocused or excluded from the electron multiplier detector.

Time-of-flight mass spectrometry and absorption spectroscopy were used by Schafer[37–40] to study the dimerization of benzyne to diphenylene in the gas phase. The benzyne was produced by the flash photolysis of benzenediazonium-2-carboxylate. It was

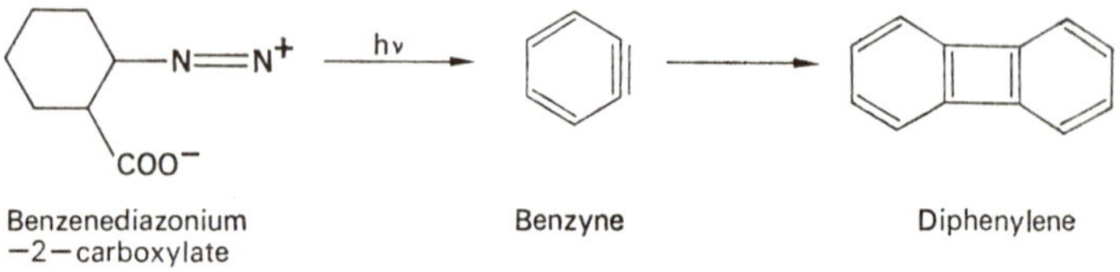

found that the species of parent mass 76 (benzyne) had the same lifetime as that of benzyne as determined by absorption spectroscopy. This confirms the ultraviolet spectrum assigned to benzyne.

R. W. Diesen *et al.*[41] coupled a flash photolysis lamp with a pyrex shock tube/time-of-flight mass spectrometer combination. The lamp was synchronized to fire at any desired time relative to the arrival of the shock wave at the end plate.[42] The technique permits photoinitiated reactions to be studied over a wide range of temperature. The conditions required for direct sampling into the mass spectrometer were evaluated using the calculations of Sherman[43] on the properties of a freely expanding jet. A special flush-end ion source was constructed to minimize the r^2 intensity loss and to eliminate electrical feedthroughs at the shock tube end of the ion source. In addition, the experimental difficulties and limitations of the technique were evaluated.

The technique was applied to the photodecomposition of nitrogen dioxide. Figure 9 shows the kinetic variation of the concentrations of the relevant species as determined by the mass spectrometer. Zero time is the arrival of the shock wave at the end plate and the indicated flash time is for the maximum flash intensity. The dashed line representing oxygen atoms is a calculated decay from the final experimental value of the rate constant for the reaction

$$O + NO_2 \longrightarrow O_2 + NO$$

An alternative and more accurate value of the oxygen atom concentration can be obtained from oxygen molecule analysis, i.e. $[O] = [O_2]_{final} - [O_2]$ (Fig. 10).

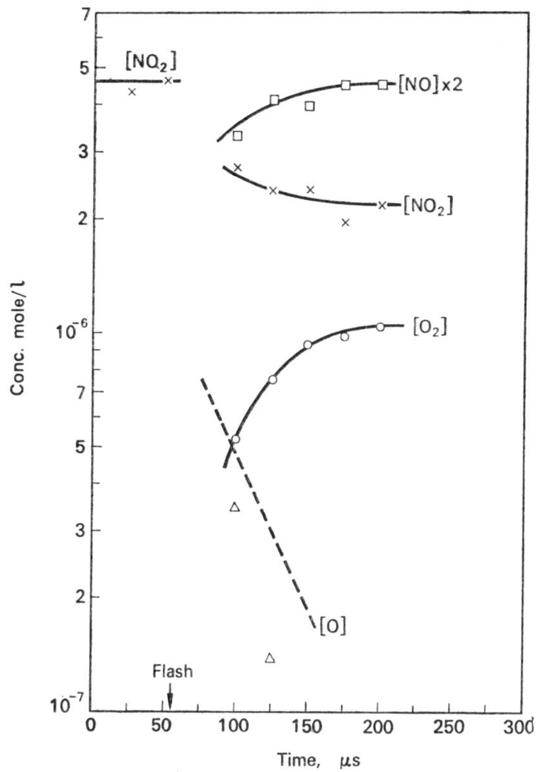

FIG. 9. Simultaneous product analysis at 1035°K. (Reprinted by kind permission of R. W. Diesen.)

The bimolecular rate constant for the reaction $O + NO_2 \rightarrow O_2 + NO$ was determined as $6{\cdot}5 \times 10^9$ l. mole^{-1} sec^{-1} at 1035°K and 2×10^9 l. mole^{-1} sec^{-1} at 298°K, giving an Arrhenius activation energy of 1·0 kcal mole^{-1}. The data compare favourably with literature values obtained at lower temperatures.

This work shows that it is feasible to monitor photoinitiated chemical reactions at high temperatures using time resolved mass spectrometry. The approach appears limited to photochemical decompositions of species with relatively high extinction coefficients.

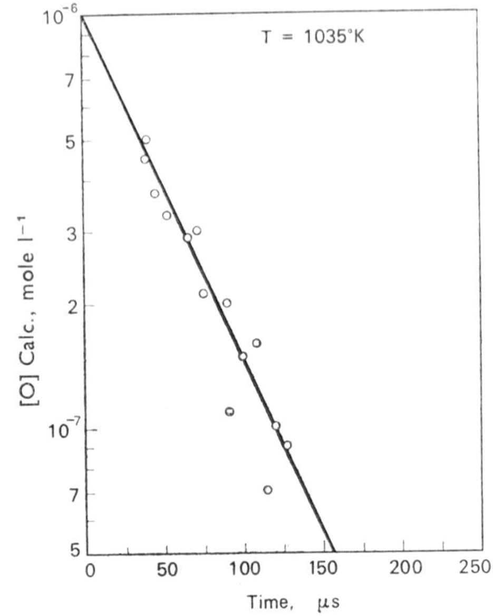

FIG. 10. Disappearance of oxygen atoms at 1035°K by the reaction $O + NO_2 \xrightarrow{k_2} O_2 + NO$. Zero time is peak of flash and [O] is calculated from O_2 analysis by $[O] = [O_2]_f - [O_2]$. A value of $k_2 = 6{\cdot}5 \times 10^9$ l. mole^{-1} sec^{-1} is calculated from $d \ln [O]/dt = k_2[NO_2]_{av}$ where $[NO_2]_{av} = 2{\cdot}9 \times 10^{-6}$ mole l^{-1}. (Reprinted by kind permission of R. W. Diesen.)

CONCLUSION

Because of the experimental and instrumental problems encountered, few papers have been published on the use of time resolved mass spectrometry to monitor flash photolysed reactions since the introduction of the technique in 1957. Improvements in experimentation made at Sandia Laboratories have shown that the technique is viable. This has encouraged several other research groups to enter the field with the aim of making the method more generally applicable to photochemical kinetics.

REFERENCES

1. (a) PORTER, G., in *Techniques in Organic Chemistry*, Edited by A. Weissberger, Vol. VIII, 2nd edn., Part II, Chapter XIX. (Wiley, New York, 1963).
(b) NORRISH, R. G. W., *Chem. Brit.* 1965, 289.

2. Allen, J. H., McKellar, J. F., and Pearce, V. F., *J. Sci. Instr.*, **40,** 372 (1963).
3. O'Callaghan, W. B., Gunning, H. E., Kebarle, P., Strausz, O. P. and Tyerman, W. J. R. *J. Am. Chem. Soc.* **88,** 4277 (1966).
4. Gunning, H. E., Kato, M., Kebarle, P., Masamune, S., Strausz, O. P. and Tyerman, W. J. R. *Chem. Comm.* 497 (1967).
5. Kistiakowsky, G. B., and Kydd, P. H., *J. Am. Chem. Soc.* **79,** 4825 (1957).
6. Norrish, R. G. W., and Porter, G., *Proc. Roy. Soc. (London)*, **A210,** 439 (1952).
7. Meyer, R. T., Paper presented at meeting of New Mexico Academy of Science, Albuquerque, New Mexico, October 1961.
8. Meyer, R. T., Paper presented at Bendix Users' Clinic, 10th Annual Meeting of ASTM Committee E-14 on Mass Spectrometry, New Orleans, La., June 1962.
9. Meyer, R. T., Paper presented at 6th Annual Bendix Time-of-Flight Mass Spectrometer Symposium, Cincinnati, Ohio, October 1964.
10. Meyer, R. T., and Freese, J. M,. Paper presented at 13th Annual Conference on Mass Spectrometry and Allied Topics, St. Louis, Mo., May 1965.
11. Meyer, R. T., *J. Sci. Instr.* **44,** 422 (1967).
12. Meyer, R. T., Paper No. 80, Division of Physical Chemistry, American Chemical Society, 151st Meeting, Pittsburgh, Pa., March 1966.
13. Meyer, R. T., *J. Chem. Phys.* **46,** 967 (1967).
14. (a) Meyer, R. T., *J. Chem. Phys.* **46,** 4146 (1967).
    (b) Freese, J. M. and Meyer, R. T., *Rev. Sci. Instr.* **39,** 1764 (1968).
15. Damoth, D. C., Paper presented at 12th Annual Conference on Mass Spectrometry and Allied Topics, Montreal, Canada, May 1964; also this book, Chapter 1.
16. Meyer, R. T., *Rev. Sci. Instr.* **35,** 1064 (1964); also this book, page 74.
17. Meyer, R. T., Olson, C. E., and Berlint, R. R., Paper presented at 14th Annual Conference on Mass Spectrometry and Allied Topics, Dallas, Texas, May 1966; also this book, page 82.
18. Meyer, R. T., Paper presented at 4th National Meeting, Society for Applied Spectroscopy, Denver, Colo., August 1965.
19. Meyer, R. T., Paper presented at 15th Annual Conference on Mass Spectrometry and Allied Topics, Denver, Colo., May 1967.
20. Meyer, R. T., Paper No. 17, Division of Physical Chemistry, American Chemical Society, 149th Meeting, Detroit, Mich., April 1965; *J. Phys. Chem.* **72,** 1583 (1968).
21. Dainton, F. S., and Norrish, R. G. W., *Proc. Roy. Soc. (London)*, **A177,** 393, 421 (1941).
22. Foord, S. G., and Norrish, R. G. W., *Proc. Roy. Soc. (London)*, **A152,** 196 (1935).
23. Ford, H., *Can. J. Chem.* **38,** 1780 (1960).
24. Kaufman, F., and Del Greco, F. P., *Ninth Symposium (International) on Combustion* (Academic Press, Inc., New York, 1963), p. 659.
25. Rebbert, R. E., and Slagg, N., *Bull. Soc. Chem. Belges* **71,** 709 (1962).
26. Brown, H. W., and Pimentel, G. C., *J. Chem. Phys.* **29,** 883 (1958).
27. Hirschlaff, E., and Norrish, R. G. W., *J. Chem. Soc.* 1580 (1936).

28. Napier, I. M., and Norrish, R. G. W., *Proc. Roy. Soc. (London)*, **A299,** 317 (1967).
29. Donovan, R. J., and Husain, D., *Trans. Faraday Soc.* **62,** 2023 (1966).
30. Porter, G., Szabo, Z. G., and Townsend, M. G., *Proc. Roy. Soc. (London)*, **A270,** 493 (1962).
31. For a review of earlier work, see Majer, J. R., and Simons, J. P., *Advances in Photochemistry* (Interscience Publishers, New York, 1964), Vol. 2, pp. 137–181.
32. Porret, D., and Goodeve, C. F., *Proc. Roy. Soc. (London)*, **A165,** 31 (1938).
33. Kasper, J. V. V., and Pimentel, G. C., *Appl. Phys. Letters* **5,** 231 (1964).
34. Kasper, J. V. V., Parker, J. H., and Pimentel, G. C., *J. Chem. Phys.* **43,** 1827 (1965).
35. Parker, J. H., and Pimentel, G. C., Paper No. 74, Division of Physical Chemistry, American Chemical Society, 152nd Meeting, New York, NY, September 1966.
36. Kasper, J. V. V., Paper No. 75, *ibid.*
37. Schafer, M., *Diss. Abs.* **27,** 444B (1966).
38. Berry, R. S., and Schafer, M. E., *J. Am. Chem. Soc.* **87,** 4497 (1965).
39. Berry, R. S., Clardy, J., and Schafer, M. E. *J. Am. Chem. Soc.* **86,** 2738 (1964).
40. Berry, R. S., Clardy, J., and Schafer, M. E., *Tetrahedron Letters*, 1003–17 (1965).
41. Diesen, R. W., Felmlee, W. J., and Petrella, R. V., The Dow Chemical Company, Midland Mich., USA. Private communication.
42. Robinson, P. A., this book.
43. Sherman, F. S., *Phys. Fluids* **8,** 773 (1965).

FLASH PYROLYSIS REACTIONS

W. K. Joy

The British Coal Utilization Research Association, Leatherhead, Surrey

INTRODUCTION

This section describes the use of a time-of-flight mass spectrometer for examining the volatile products obtained from the flash pyrolysis of solids. Only six authors have published in this comparatively new field: two of these describe flash-tube irradiation of solids, three pyrolysis using a ruby-laser beam, and one a pulse heating technique for thin metal wires.

When a solid receives a pulse of intense radiation from, for example, a flash tube, energy is absorbed thus raising the temperature of the substance. The amount of energy absorbed varies according to the absorption characteristics of the substance: black substances absorb more energy than white whilst the energy absorbed by substances normally regarded as transparent is very small. Transparent and light-coloured substances can be heated indirectly by incorporating a small amount of an inert black substance, such as carbon black. The heat from this will be transferred to the non-absorbing substance. This method is analogous to the addition of a small amount of a photosensitive gas to an inactive gas in flash photolysis.

Light from a xenon flash-lamp can be focused on to a sample of a few milligrams of powdered material by means of reflectors, the flash-lamp often being in the form of a spiral or coil and the sample being on the walls of a tube laid along the axis. However, it is not possible to focus a substantial amount of the light produced by a flash-tube on to the very small area of the powdered sample which could be evaporated in the source of the time-of-flight mass spectrometer. Therefore in the flash heating experiments described, larger samples have been heated externally to the source.

Some of the disadvantages of flash-heating can be overcome

by the use of a laser. It is relatively a simple matter to focus a laser beam on to a small sample in the source of a time-of-flight mass spectrometer. The laser can be situated outside the vacuum system of the mass spectrometer and the beam can enter through a quartz window. The pyrolysis must necessarily be carried out *in vacuo*. The sample can be placed in the source a few millimetres from the ionizing beam without the need for any slit or orifice intervening which might cause the loss of transient species. In high temperature pyrolysis, radicals and ions are often produced in the volatiles and it is possible to detect them with the mass spectrometer. Very small samples, as little as a few micrograms, can be heated and the difficulty of having a large amount of gas evolved in the source, which affects the operation of the mass spectrometer, is avoided.

If the sample is in lump form only part of it will be heated so that the products will originate mainly from the very hot part of the surface which will probably be evaporated leaving a crater but also from the surrounding solid which has not been heated to such a high temperature. The best form for the solid from the point of view of reaching a uniform high temperature is that of fine particles: metals in the form of thin foils and organic substances such as cellulose have also proved to be effective.[(1)]

FLASH-TUBE PYROLYSIS

This was investigated by Lincoln and Friedman; the design of inlet system devised by Lincoln[(2)] is illustrated in Fig. 1. The sample is contained in a small tube placed along the axis of the helical flash tube and in direct line with the electron beam in the source of the time of flight mass spectrometer. When the tube is fired the products are drawn rapidly into the source and analysed. The spectra are displayed on two oscilloscopes.[(3)] One is used to follow the change of the peak height at one preselected m/e value for 500 msec during a pyrolysis, while the other oscilloscope displays a complete spectrum in which the horizontal axis represents m/e, the vertical axis represents time and the intensity of the display at any point represents ion abundance. The displays are recorded on 'Polaroid' film; Figures 2 and 3 illustrate the mass spectra of the products from flash-heated glycolaldehyde. It can be seen from both displays that mass 52 quickly decays while mass 61 rises to a maximum.

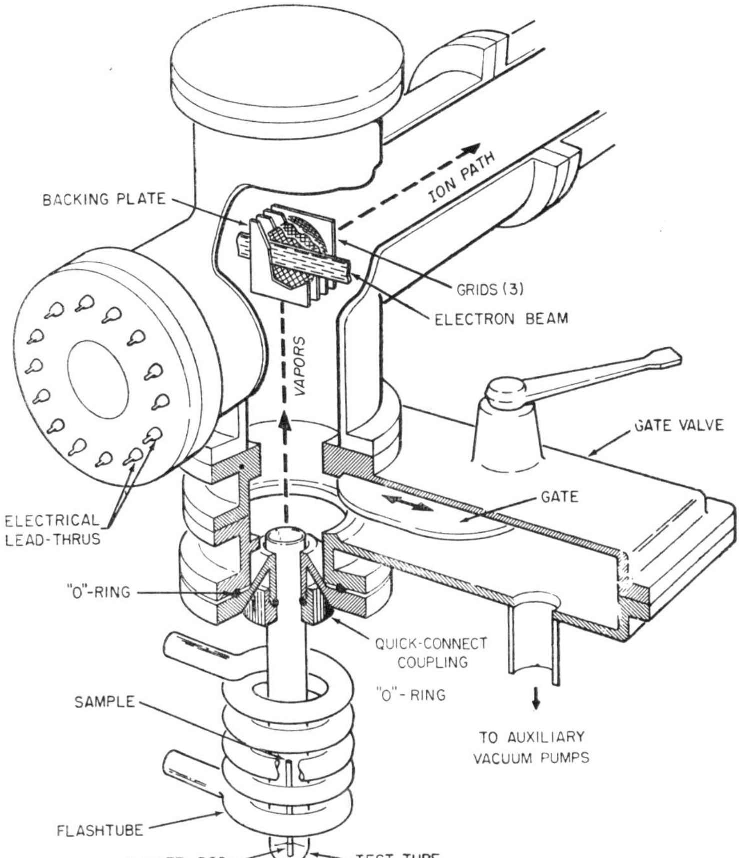

FIG. 1. Reprinted from *Anal. Chem.* **37**, 542 (1965). (Copyright (1965) by the American Chemical Society. Reprinted by kind permission of the copyright owner and K. A. Lincoln.)

Lincoln was interested in the production of inflammable gases from solid fuels under various intensities of radiation, as possible fire hazards. He chose cellulose as a representative type of material in the form of thin sheets incorporating 2% carbon black. The fact that negligible amounts of char and tar were produced in the flash-heating experiments, as compared with slower rates of heating, was taken as an indication that the cellulose was completely evaporated. The production of volatile organic material and an increase in the CO/CO_2 ratio with higher fluxes is significant to any study of fire hazards. Short-lived species were reported as occurring at high radiation fluxes but these were not identified, and the author does not give details of the evidence.

Friedman[4] has examined the products from the flash pyrolysis of a phenolformaldehyde resin. The samples were placed along the axis of a coiled flash lamp which was filled with xenon to a pressure of 40 mmHg. A 220 μF capacitor bank at 5000 V was discharged through the flash-tube. The pyrolysis products flowed through a 2 μm hole in a gold foil disc into the mass spectrometer source. Spectra were recorded by means of two analogue scanners which were triggered, after a predetermined time delay, and set to scan the mass ranges 1–100 and 12–130; the sweep time was 7 sec. The output signals from the scanners were fed to an oscilloscope set in the single sweep mode and triggered 3 sec after the flash; the mass spectra were recorded with a Polaroid camera. Each sample was subjected to eleven flashes, time being allowed between successive flashes for the reaction products to be pumped away.

Examination of the mass spectra of the products showed hydrocarbons up to C_8, H_2, CO, CO_2 and H_2O. The products showed two different types of variation with successive flashes; H_2, CO, CH_4 and C_4 and heavier hydrocarbons showed an increase to a maximum at the fourth flash while H_2O, CO_2 and the C_2 and C_3 hydrocarbons showed a continual decrease.

A measure of the surface temperature of the resin during pyrolysis was obtained from the water gas equilibrium:

$$K = \frac{(H_2O)\,(CO)}{(CO_2)\,(H_2)}$$

It was assumed that equilibrium on the surface was reached and that the products were then rapidly quenched on leaving it. The temperatures calculated showed a decrease from about 2000°K for the first flash to about 1400°K for the last; this decrease is attributed to the collection of carbon on the flash tube which would reduce its efficiency. In view of the results of experiments by Zavitsanos, mentioned later, the attainment of equilibrium is doubtful.

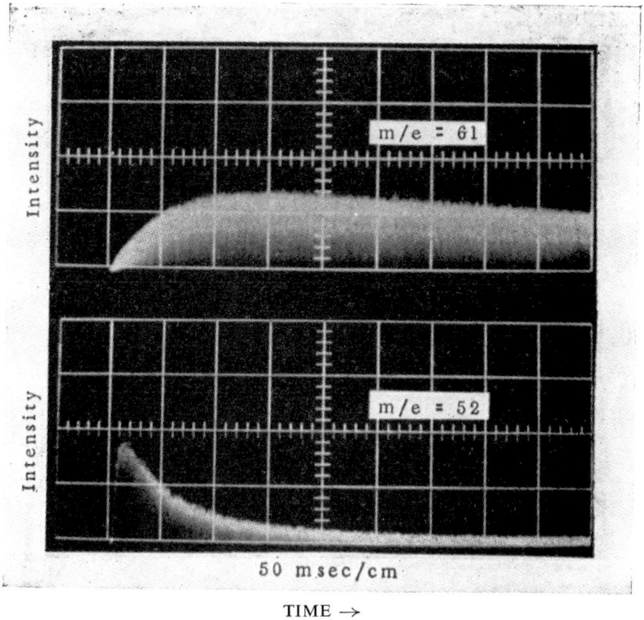

FIG. 2. Typical time-resolved raster display. (Reproduced from *Rev. Sci. Inst.* **35,** 1690 (1964), by kind permission of K. A. Lincoln and the American Institute of Physics.)

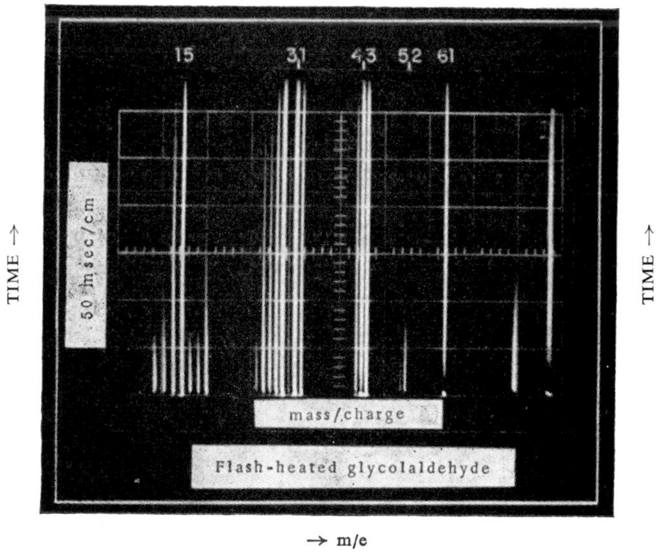

FIG. 3. Peak height vs. time of two glycolaldehyde peaks. (Reproduced from *Rev. Sci. Inst.* 35, 1690 (1964), by kind permission of K. A. Lincoln and the American Institute of Physics.)

LASER PYROLYSIS

Vastola and Pirone[5] have published the results of preliminary experiments in which they used a laser beam in the source of a time-of-flight mass spectrometer to heat small lump samples of coal. By fine focusing of the beam, individual coal macerals at the surface could be vaporized. The apparatus is illustrated in Fig. 4. The beam from a ruby laser passed by means of a right angle prism down through a focusing lens and through a window

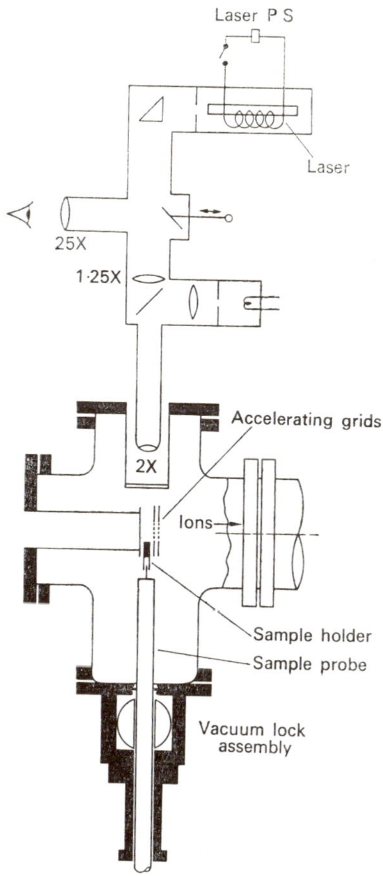

FIG. 4. *Chem. Eng. News*, 24 Oct. 1966, 49. (Reprinted by kind permission of the Editor of *Chemical and Engineering News* and F. J. Vastola.)

into the source of the time-of-flight mass spectrometer. The sample was situated below the ion grids and the laser beam passed through the path of the electron beam to reach it; thus the products were evolved just below the electron beam. A microscope attachment was used for observing the sample and for focusing the beam on to it.

A mass spectrum was recorded on Polaroid film from the oscilloscope operating in the "single shot" mode at different delay times after firing the laser. The results showed that aromatic and aliphatic substituted hydrocarbons such as benzene, toluene, xylenes, naphthalene and methyl- and dimethyl-naphthalenes were present. An unusual distribution of peaks in the cracking patterns at high mass numbers suggested that some hydrogen deficient species of aromatic hydrocarbons were produced by the laser heating.(6) These peaks are attributed to the action of radicals and hydrocarbons in the source before electron bombardment. Vastola and Pirone concluded that the system is capable of detecting primary coal pyrolysis products.

An apparatus similar to that used by Vastola and Pirone has been built at the British Coal Utilization Research Association to study the initial products from the rapid heating of coal particles. Finely divided coal ($< 10\ \mu m$) was placed on a quartz support in the source of a time-of-flight mass spectrometer and heated from above with a ruby laser. The spectra were recorded from an oscilloscope display on 70 mm film at 200 μsec intervals by means of a high speed drum camera.(7) Preliminary results(7) show that at very high temperatures, besides acetylene, methane, water vapour and smaller quantities of C_4 and C_5 hydrocarbons, unstable species such as methyl radicals, carbon and hydrogen atoms are produced. Also mass peaks at 50, 74 and 98 indicate the series of polyacetylenes, C_4H_2, C_6H_2 and C_8H_2.

Zavitsanos(8) has studied the evaporation of graphite using a ruby laser. The sample was held in the source of a time-of-flight mass spectrometer about 3 cm from the ionizing electron beam. The laser beam was focused on to the specimen at an angle to its surface. Separate experiments without the mass spectrometer showed that in both air and vacuum the products of vaporization were ejected normally from the surface of the sample irrespective of the direction of the beam. Experiments using a silica target indicated that nearly all of the available energy was used in evaporation of the silica.

Mass spectra of the evaporated species from laser-heated graphite were obtained by photographing an oscilloscope set in the "single-sweep" mode. By varying the time between firing the laser and triggering the oscilloscope changes in the mass spectrum with time could be observed.

Several species were found in the vapours, from C_1 up to C_{10}, and the variation of proportions of these with time were recorded. It was shown that equilibrium was not obtained, since calculations made from the heats of vaporization of individual species at 4100°K did not agree with the composition of the vapour found by mass spectrometry.

The rates of vaporization of several graphites for various laser energies were calculated from the radiant energy and the mass evaporated.

FLASH PYROLYSIS OF METALS AND METALLIC COMPOUNDS BY CONDUCTION HEATING

In a recent paper(9) a method of studying the vapour pressure of metals and their oxides and nitrides at high temperature is described. Short pieces of very fine wire were inserted into a Bendix 843A probe, being held by screws in sapphire insulators. Oxides and nitrides were prepared by electrically heating the wires in air. When the probe was inserted into the source, samples thus treated and also samples of pure metal wire could be heated by electrical discharge through them; they were only 6 mm from the source grids. The amount of oxidation or nitridation was estimated from the change in resistance of the wire.

On heating in the source the mass spectra of the resulting products were recorded on a drum camera. The vaporization of metals could be followed in time and the liberation of products such as ZrO and ZrO_2 from oxidized zirconium identified. Liberation of hydrocarbons from the metal surface on first heating masked any products liberated at lower masses, but by successively flashing the filament this could be kept down to a convenient level in subsequent experiments.

REFERENCES

1. Nelson, L. S., *Science* **136,** No. 3513, 296–303 (1962).
2. Lincoln, K., *Pyrodynamics* **2,** 133–43 (1965).
3. Lincoln, K., *Rev. Sci. Instr.* **35,** 1688–91 (1964).
4. Friedman, H. L., *J. Appl. Poly. Sci.* **9,** 651–62 (1965).
5. Vastola, F. J., and Pirone, A. J., Symposium on Pyrolysis Reactions of Fossil Fuels, Pittsburgh, March 1966.
6. Vastola, F. J., private communication.
7. Joy, W. K., Ladner, W. R., and Pritchard, E. *Nature* **217,** 640–1 (1968).
8. Zavitsanos, P. D., General Electric, Technical Information Series R67SD11, 28 pp., 1967.
9. Meyer, R. T., and Ames, L. L., in *Applications of Mass spectrometry in Inorganic Chemistry*, Edited by J. L. Margrave, American Chemical Society Advances in Chemistry Series, 1967.

SHOCK WAVE REACTIONS

P. A. ROBINSON†

Department of Physical Chemistry, University of Cambridge

THE shock tube is now an established tool for producing high temperatures, homogeneously, in the gas phase. However, the total analysis time may be only a few milliseconds and it is desirable to obtain as much information as possible in one experiment. The time-of-flight mass spectrometer is suited to shock-tube analysis since it is able to produce a series of mass spectra throughout the available observation time. The working pressure of a mass spectrometer is of the order of 10^{-9} of that existing in the shock-heated gas. This communication reviews some of the work that has been done concerning the problem of coupling the shock tube to a mass spectrometer.

Two methods of coupling have been developed. The first of these employs a pin-hole leak or small nozzle in the end wall of the shock tube. The second and more recent technique consists of expanding the shock-heated gas through a nozzle system to create a molecular beam, which then enters the mass spectrometer.

COUPLING BY MEANS OF A PIN-HOLE LEAK OR SMALL NOZZLE IN THE END WALL OF THE SHOCK TUBE

With this technique, the gas sample that is analysed is processed by both the incident and reflected shock waves. Upon reflection at the end wall, the shock wave brings about a further increase in temperature of the experimental gas, which becomes stagnant. Sampling into the mass spectrometer is then by effusion. Analysis time is terminated by the reflected shock wave interacting with the contact surface and returning to the end wall.

† Address now: Petrochemical and Polymer Laboratory, I.C.I. Ltd., P. O. Box II, Runcorn, Cheshire, England.

This method of sampling was originally employed by Bradley and Kistiakowsky.[1-3] The apparatus was modified by Dove and Moulton[4] although the principle remained the same. In the early work,[1] the sampling leak consisted of thick gold foil with a 0·005 in. diameter hole in the centre. Later the leak was in the form of a small cone, the aperture at the apex being 0·08 mm diameter. This was done to reduce the portion of gas being drawn from the thermal boundary layer, formed on the end wall of the shock tube.

Modifications were made to the standard time-of-flight mass spectrometer in adapting it to shock tube analysis.[4] The pumping speed of the ion source region was increased and a large ballast volume was added to reduce the rate at which the pressure rose after the arrival of the shock wave at the leak. The ion source region and the drift tube were connected by an opening of 1 cm^2 so that the pressure in the drift tube (pumped separately to the ion source) was low (10^{-8} mm) during the observation period. The filament and high voltages were automatically cut off when the pressure reached a predetermined value.

In the work of Bradley and Kistiakowsky successive oscilloscope sweeps, triggered by the accelerating pulses, displayed complete mass spectra which were photographically recorded by a rotating drum camera. Dove and Moulton avoided the use of moving film by obtaining a series of mass spectra at 50 μsec interval, displayed one below the other on the oscilloscope screen. This was achieved by adding algebraically the mass spectrum signals and a saw-tooth wave form of duration equal to 5–8 cycles of the mass spectrometer. Continuing the work, Gay *et al.* have extended the technique to give spectra at intervals, variable between 15 and 40 μsec.[5] Dove remains active in the field with a shock tube constructed at Toronto.[6]

Two sampling techniques, one involving a simple pin-hole and the other a slender pyrex nozzle protruding into the reflected shock region, have been studied by Modica.[7] By studying the kinetics of the nitrous oxide decomposition, he was able to show that below reflected shock temperatures of 3000°K the two methods of sampling gave the same results. Above this temperature, however, the effects of the cold end wall on the hot, "shocked" gas introduced a systematic error in the measured rate constant: the error was less for the nozzle sampling.

An experimental approach similar to that of Bradley and

Kistiakowsky was independently developed by Diesen and Felmlee.[8]

Gutmann *et al.*[9] coupled a quadrupole mass spectrometer to a shock tube by means of a conical orifice, 0·002 in. diameter. Their results showed that two important considerations in the design of the equipment were an extremely rapidly pumped ion source and a microscopically smooth coupling orifice.

The problem of removing the sampled gas from the ion source at the necessary speed has been considered recently by Ryason.[10] His leak was formed from a stainless steel disc with a centrally located hole (0·1 mm), the shape of which was a short cylindrical section leading to a 90° conical section. His results indicated a rapid build up of pressure using the Bradley and Kistiakowsky configuration. Adequate pumping was obtained by placing the shock tube and flight tube axis perpendicularly, with the sample beam passing through the ionizing electron beam into a cryosorption pump. This arrangement proved satisfactory and results on the decomposition of carbon disulphide were in reasonable agreement with those of other workers.

The following possible errors must be considered when adopting this simple method of coupling.

Non-ideal behaviour of reflected shock waves

Strong shock waves do not reflect as plane waves, but instead the reflected wave divides near the wall into two or more shock waves which do not move with uniform velocity. This may be avoided by dilution with noble gases.

Reflected shock waves in pure noble gases and mixtures of these with a small percentage of non-ideal gas have a lower velocity than predicted from the incident shock velocity. This arises from the interaction of the reflected shock wave and the boundary layer formed behind the incident shock wave. However, sampling the gas on the centre axis, well away from the boundary layer suggests that this attains the predicted reflected shock temperature.

Background scattering and pressure build-up inside the mass spectrometer

Kydd[11] has shown that the increase in background pressure is small compared with the effective pressure of the sample being analysed by the electron beam when it is close to the leak. Hydrodynamic flow through the leak assures that only a small fraction of the sample will collide with the surface. The use of ballast volumes and fast pumped sources have been already noted. Internal collisions in the mass spectrometer must also be minimized.

End wall effects

A boundary layer of cool gas grows from the end wall and the interference of this on the integrity of the sample is difficult to assess. Bradley and Kistiakowsky estimated the effect to be small below 1800°K. The surface also allows an opportunity for catalytic recombination of atoms and radicals to occur. The relative influence of the end wall will increase with increasing temperature and decreasing pressure. This was observed in the results of Diesen and Felmlee, and Modica.

Marsters *et al.*[12] have analysed the effect of the boundary layer. They found that the volume that fed the sampling hole grew linearly with time, regardless of whether free molecular or hydrodynamic flow was obtained. The radius of this hemisphere, centred at the sampling hole, therefore increased as $r \propto t^{\frac{1}{3}}$.

The rate at which the thermal boundary layer diffused into the stagnated gas was $d \propto t^{\frac{1}{2}}$.

The thermal boundary layer thickness quickly exceeded the hemisphere radius and only the very early portion of the flow was free from end wall effects.

Bradley and Kistiakowsky argued that in the case of hydrodynamic flow the boundary layer may be eliminated from the end wall near the leak. Sturtevant,[13] studying ionization reactions in gases behind reflected shock waves, showed that primarily because only the initial stages of the reaction were studied and also because the reactive species were in small concentrations, it was unlikely that the sample was perturbed by the thermal layer.

EXPANDED FLOW TECHNIQUE

This method of coupling a mass spectrometer to a shock tube circumvents the uncertainties surrounding the technique of sampling through a pin-hole in the end wall, and also enables a study to be made of the gas heated by the incident shock wave.

Expansion of the shock-heated gas through a nozzle system has been developed and studied in recent years for the generation of molecular beams, in particular the work of Skinner.[14] Assuming ideal gas behaviour, quasi-one dimensional and isentropic flow during the expansion the change in pressure, density, temperature and Mach number can be evaluated. These calculations indicate a rapid initial rate of decrease of temperature (10^{9}°K per sec) and density which causes a "freezing" of the gas composition and the sample integrity to be maintained. This cooling effect can be visualized as a transformation of the random thermal motions of the shock-heated gas to directed flow through the expansion. The nozzle system usually takes the form of an expansion cone followed by a skimmer and then collimating orifices (Fig. 1). The

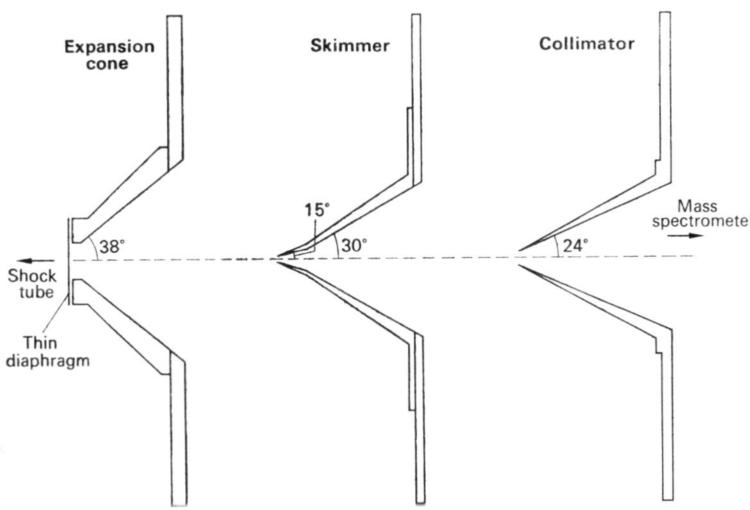

FIG. 1. Nozzle system.

skimmer faces the supersonic flow from the first orifice and so has to be designed to minimize the formation of a detached shock wave across the skimmer opening. The transition from continuum flow to free molecular flow takes place downstream of the skimmer. Pressures and dimensions must be such that scattering is minimized.

The expansion system must be separated from the experimental section of the shock tube by a very thin diaphragm which will rupture easily upon the arrival of the shock front. A combination of films formed from two varieties of "dope" has been found to produce a satisfactory diaphragm.[15]

Recently, Marsters *et al.* have constructed an elegant apparatus, which is described in ref. 12. For the design of the collimating system they were guided by recent work of Skinner.[14] Cryosorption pumping proved effective for removing large quantities of gas in times of the order of milliseconds. The second and remaining nozzles were offset by 7° to avoid diaphragm dust. A very thin collodion diaphragm was positioned at the entrance to the first nozzle.

This group was concerned with low flow Mach numbers behind the shock front and so the gas sample was allowed to expand freely until the flow reached a sufficiently high Mach number to permit the supersonic nozzle to operate satisfactorily (with an attached shock wave on the nozzle).

They have analysed the initial expansion process occurring between the end of the shock tube and the nozzle. Results showed that the rate of cooling of the gas was in general lower for the lower temperature flows and the time spent in the expansion was considerably longer. In order to increase the cooling rate and decrease the time in the expansion Marsters *et al.* varied the radius of the exit, this being the only parameter which could be adjusted. Hence, a sleeve was introduced into the shock tube and so the running time of the experiment became limited by the time taken by the shock, reflected from the closed annular area, to reach the opening of the sleeve.

A further consequence of the sleeved operation was the possibility of choking of the flow as a result of boundary layers set up inside the sleeve. Marsters *et al.* showed that a sleeve should be avoided at low initial pressures and low Mach numbers. At high temperatures and high initial pressures, they suggested that the sleeve method should prove satisfactory.

A problem relating to sample integrity is introduced by the nature of the expanded flow technique since mass separation in supersonic molecular beams is well known. Becker *et al.*[16] demonstrated that heavier components were enriched relative to the lighter along the axis of the beam. Recent work by Andres *et al.*[17] suggested that the separation was not due entirely to radial migration but may be also a consequence of skimmer-jet interactions in these flows. Work is being continued to characterize the nature of these interactions and to eventually eliminate them.

Marsters *et al.* are aware of such effects and are also tackling the problem.

REFERENCES

1. BRADLEY, J. N., and KISTIAKOWSKY, G. B. *J. Chem. Phys.* **35,** 1, 250 (1961).
2. BRADLEY, J. N., and KISTIAKOWSKY, G. B., *J. Chem. Phys.* **35,** 1, 264 (1961).
3. BRADLEY, J. N., *Trans. Faraday Soc.* **57** (10), 1750 (1961).
4. DOVE, J. E., and MOULTON, D. MCL., *Proc. Roy. Soc.* **A283,** 216 (1965).
5. GAY, I. D., KERN, R. D., KISTIAKOWSKY, G. B., and NIKI, H., *J. Chem. Phys.* **45,** 2371 (1966).
6. DOVE, J. E., CRESWELL, R., and DI VALENTIN, M. A., *Phys. of Fluids* **9,** 2285 (1966).
7. MODICA, A. P., *J. Phys. Chem.* **69,** 2111 (1965).
8. DIESEN, R. W., and FELMLEE, W. J., *J. Chem. Phys.* **39,** 2115 (1963).
9. GUTMANN, D., HAY, A. J., and BELFORD, R. L., *J. Phys. Chem.* **70,** 1786 (1966).
10. RYASON, P. R., *Rev. Sci. Instr.* **38** (5), 607 (1967).
11. KISTIAKOWSKY, G. B., and KYDD, P. M., *J. Am. Chem. Soc.* **79,** 4825 (1957).
12. MARSTERS, G. F., BAUER, S. H., and RESLER, E. L., *5th International Symposium on Shock Tubes Maryland, USA, 1965.*
13. STURTEVANT, B., *J. Fluid Mech.* **25** (4), 641 (1966).
14. SKINNER, G. T., and MOYZIS, J., *Phys. of Fluids* **8,** 452 (1965).
15. BRADLEY, J. N., and ROBINSON, P. A., *Proc. Roy. Soc. (London)* **A301,** 285 (1967).
16. BECKER, E. W., BIER, K., and BURGHOFF, H., *Z. Naturf.* **10a,** 565 (1955).
17. ANDRES, R. P., ANDERSON, J. B., and FENN, J. B., *Adv. Chem. Phys.* **10,** 275 (1966).

FAST RECORDING TECHNIQUES FOR TIME RESOLVED MASS SPECTROMETRY†

R. T. MEYER

Sandia Laboratories, Albuquerque, New Mexico, USA

INTRODUCTION

The application of the time-of-flight mass spectrometer to kinetic studies of fast reactions has been possible only through the parallel development of specialized data recording methods. In a typical single experiment on a transient phenomenon, the spectrometer provides thousands of data correlating mass number and ion intensity as a function of time. While the experiment could, in some applications, be repeated many times to obtain a separate ion intensity versus time record for each mass number, or an ion intensity versus mass record for separate selected points in time, the tendency has been to utilize the full capacity of the spectrometer and to obtain the complete three-dimensional array of data for each experiment. This distinctive feature of the time-of-flight mass spectrometer permits the retrieval of a self-consistent set of data for every individual experiment. A further advantage would exist in the form of an increased signal-to-noise ratio if the transient experiment could be repeated with the ion intensities being successively added for identical mass and time coordinates. However, many experimental systems, e.g. gas phase flash photolysis and shock-initiated gaseous decompositions, are not conservative in the reactants. These systems are difficult to reproduce as exactly and as rapidly as required for an efficient integration of data.

† This work was supported by the United States Atomic Energy Commission.

Several two- and three-dimensional data recording techniques using oscilloscopes and cameras are presently available. Although these methods were developed by different investigators for a variety of time resolved mass spectrometry applications, all are suitable to any problem. The principal criteria were the ease of data acquisition with available instrumentation and the number of data required for a specific kinetic analysis. No single technique has been widely adopted. This paper will describe equitably the several techniques and evaluate their advantages and disadvantages. The order of the discussion reflects only the author's view of the increase in quantitative data that can be retrieved and should not be considered overly significant. In addition, the discussion describes briefly some data recording and processing techniques which are either under development or need to be explored in order to improve the efficiency and accuracy of data reduction and interpretation. These techniques include magnetic tape recording, analog-to-digital conversion, and computer processing.

OSCILLOSCOPE DATA RECORDING TECHNIQUES

Figure 1 is a schematic summary of the five principal oscilloscope data recording techniques. Figure 1a is a typical ion intensity versus mass display observed with an oscilloscope readout from the electron multiplier of a time-of-flight mass spectrometer. All of the methods utilize Polaroid film recording of some version of an oscilloscope display, except for the revolving drum camera method which uses 35 mm film. Polaroid film offers the distinct advantage of rapid processing and quick determination of the validity of a given experiment.

Synchronized z-axis gating

The *z*-axis gating method was developed by Damoth.[1, 2] It is based upon intensification of the oscilloscope electron beam for the time-duration that the beam sweeps out one mass peak of each cycle of spectrometer output. By reducing the average electron beam current (oscilloscope) to the point where the other mass peaks are no longer visible on the phosphor, only the in-

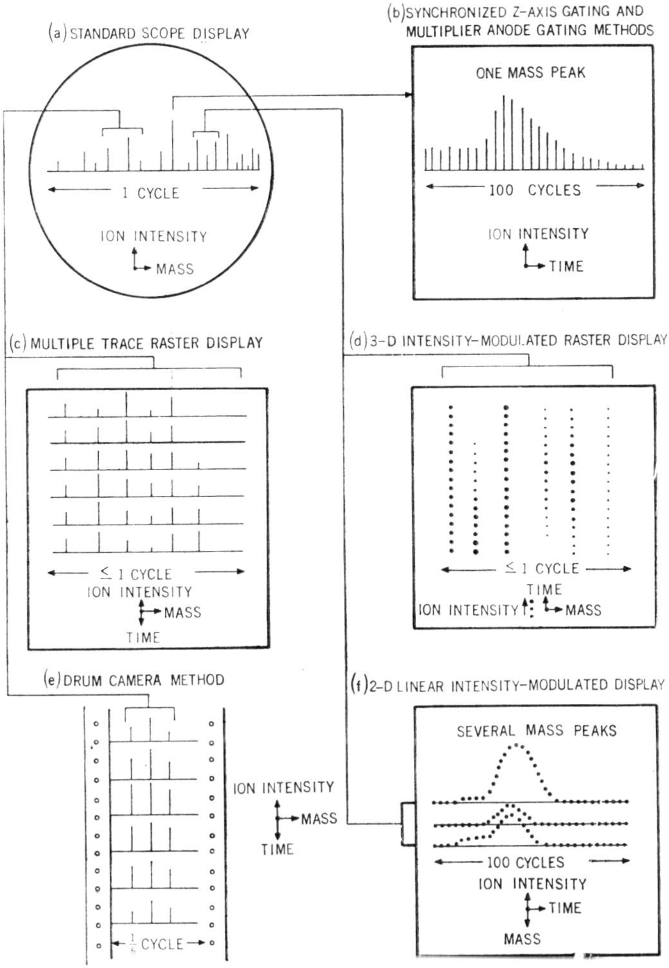

FIG. 1. Schematic summary of time resolved data display and recording by oscilloscope techniques.

tensified mass peak remains detectable. Then by decreasing the horizontal sweep rate from one cycle displayed per sweep to many cycles per sweep (0·5 msec/cm for 100 cycles at 20 kHz spectrometer frequency), a compressed display of ion intensity (y-axis) versus time (x-axis) for a single mass may be photographed.

Figure 2 shows an example of the simple circuitry required. This circuit applies a negative voltage pulse to the cathode of the oscilloscope cathode-ray-tube (CRT). The pulse is synchronized to the desired mass peak and, hence, intensity modulates the electron beam. For a Bendix spectrometer equipped with analog scanners it is most convenient to utilize the analog gate marker pulse (−150 V) as the primary signal. It is internally synchronized

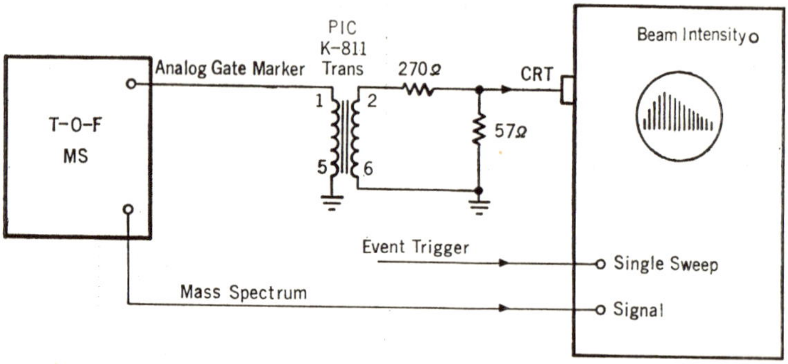

FIG. 2. Circuitry for synchronized *z*-axis intensification of selected masses. (Developed by D. C. Damoth, Bendix Corporation.)

to the spectrometer master pulse generator and needs only to be reduced in voltage through a high frequency transformer and voltage divider. The analog scanner has a built-in variable time delay for the gate marker, so the marker can be applied to any mass peak in the spectrum. The oscilloscope is operated in a single-sweep mode, which is triggered separately by a signal synchronized to the initiation of the transient event. By starting the single-sweep several cycles prior to the event, the initial ion intensity can be recorded. A baseline is established by shifting the modulating pulse off the mass peak to an adjacent no-mass (zero ion intensity) position and triggering the horizontal sweep.

The use of dual-beam oscilloscopes and multiple gate marker circuits allows one to record ion intensity versus time for several masses simultaneously. Figure 3 shows the ^{84}Kr and I_2 signals in the flash photolysis of methyl iodide in the presence of nitric oxide.[(3)]

Limitations of the synchronized *z*-axis gating method are as follows. In some transient experiments, interferences with the stability of the spectrometer electron and ion guns occur which lead to significant changes in the shape of the mass signals and to shifts in their positions on the mass scale. Thus, a *z*-axis gate which

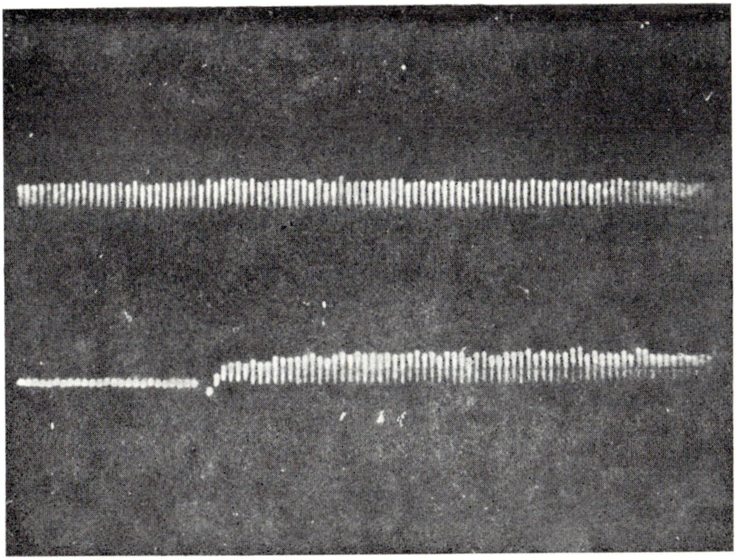

FIG. 3. Dual-beam z-axis intensification of ^{84}Kr (upper, 0·1 V/cm) and I_2 (lower, 0·05 V/cm, 0·5 msec/cm) for flash photolysis of CH_3I+ $+NO+0{\cdot}1$ Kr+25 He at 32 torr. Flash occurs at $\sim$1·25 msec with duration of $\sim$12 μsec. Constant ^{84}Kr signal indicates absence of flash heating, while growth of I_2 signal indicates termolecular recombination of I atoms (ref. 3).

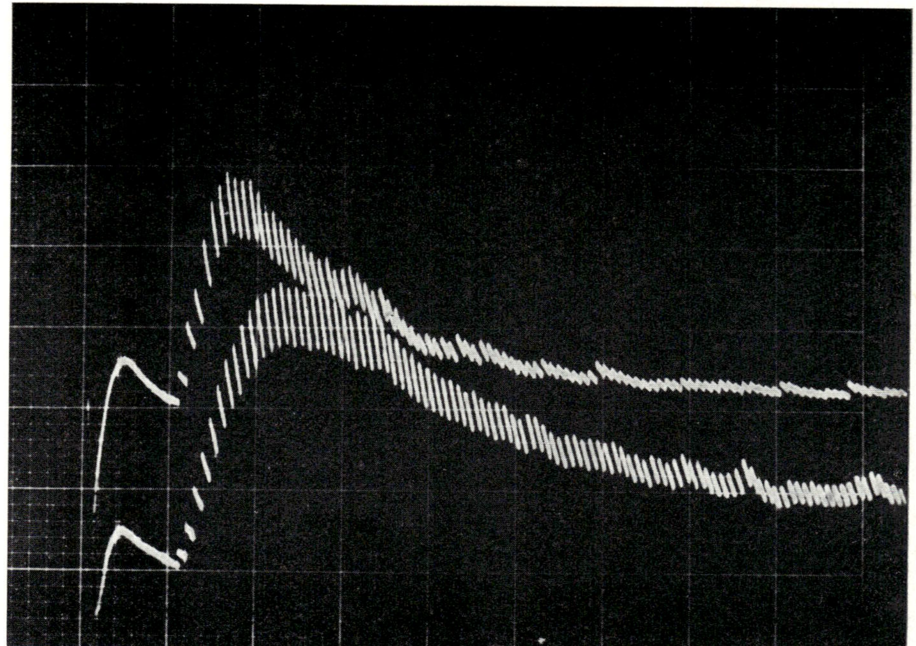

FIG. 4. Dual-trace multiplier-anode-gating for laser-induced vaporization of cellulose; $m/e = 19$ (upper) and 58 (lower). Horizontal sweep = 0·5 msec/cm. (Reproduced by permission of K. A. Lincoln; ref. 4.)

was present on a given mass peak might not remain synchronized to that peak during portions of the transient event. Even in the absence of interferences, the precise positioning of the gate on masses for intermediate species such as atoms, free radicals, and other molecular fragments may be extremely difficult. It can be achieved if a stable molecule can be employed to provide the desired mass positions. The technique also requires a separate oscilloscope electron beam for each mass recorded. On the other hand, the method is relatively simple to install and provides rapid data retrieval.

Synchronized multiplier anode gating

A technique very similar to the *z*-axis gating method was recently reported by Lincoln.[4] Instead of using the analog gate markers to modulate the oscilloscope electron beam, they are applied directly to the anode gating electrodes in the electron multiplier. This connection is exactly the same as used in the standard analog recorder operation. The time delays for the analog gate pulses are set to coincide with the desired mass signals. The multiplier anode outputs are fed directly to the vertical inputs of dual-trace or dual-beam oscilloscopes. A single horizontal sweep of the oscilloscope is triggered in the same manner as in the *z*-axis gating method. Figure 4 shows an ion intensity versus time display for two masses. The transient experiment was a laser-induced vaporization of cellulose.

Lincoln used the Tektronix model 1A1 dual-trace preamplifier with a model 547 oscilloscope. This combination permits simultaneous recording of signals from two multiplier anodes with a single electron beam. The 1A1 preamplifier permits chopping successive 500 nsec segments of each input channel at a 1 MHz rate. Since the spectrometer operates at a repetition rate of 100 kHz or less, there is no loss of information. The RC time constant for the oscilloscope input is much longer than 500 nsec and is comparable with 50 μsec for the record of Fig. 4. Thus, the electron beam traces out a succession of slanted bars corresponding to the charge and discharge of the capacitance from the mass spectral signals. Since an electron multiplier is available with six anodes, using either three dual-trace or two dual-beam oscilloscopes permits six masses to be recorded simultaneously.

Lincoln pointed out two important advantages of this method over the *z*-axis gating method: (1) the total ion current for the individual mass signal is measured rather than the peak height only, and (2) by an appropriate use of oscilloscope-input filter capacitors, the ion beam statistical fluctuations can be filtered out to produce a continuous line recording.

Three-dimensional intensity-modulated raster

This display system was devised by Lincoln to provide ion intensity as a function of time for an entire mass spectrum on one oscillogram per event.[4, 5] In order to obtain the third dimension on a single oscillogram, however, the quantitative record of ion intensity had to be sacrified for only a qualitative record. A recent version of Lincoln's system is shown schematically in Fig. 5. It is similar to *z*-axis gating in that electron beam intensification is the basis of the technique. The display coordinates were changed to provide mass along the horizontal *x*-axis, time along the vertical *y*-axis, and ion intensity via the

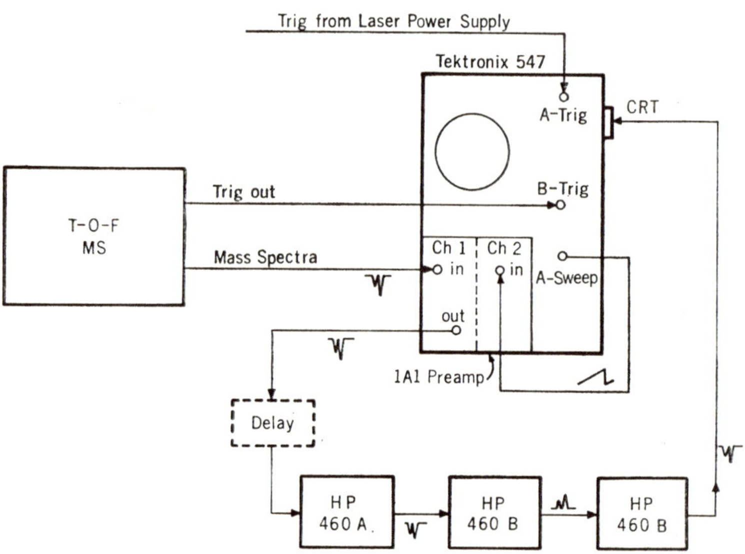

FIG. 5. Block diagram for three-dimensional intensity-modulated raster display system. (Reproduced by permission of K. A. Lincoln; ref. 4.)

density of the electron beam spot (z-axis). In this method, all of the mass peaks are intensity-modulated, and the horizontal sweep is adjusted to display one cycle or less. In addition a saw-tooth voltage is supplied to the vertical input to obtain a raster sweep.

The mass spectral signals are supplied to an external amplifier network through the vertical input and output terminals of channel 1 of a Tektronix 1A1 preamplifier. Channel 1 normally serves only as an amplification stage, except that it may be switched into the oscilloscope to provide a standard oscilloscope display. The external network amplifies all the mass spectral signals. The negative pulses are applied to the cathode of the CRT to obtain z-axis intensification for all mass positions. The degree of intensification is directly proportional to the amplitude of the mass signals, i.e. an intense mass signal generates a more negative pulse than a weak mass signal. The larger negative pulse, when applied to the cathode, produces a larger electron beam current and a more intense phosphorescence. Thus, the intensity of the phosphorescence is a measure of the amplitude of the mass signal.

The time base B horizontal sweep of the Tektronix 547 oscilloscope is triggered at the frequency of the spectrometer trigger signals which produces a horizontal display of dots across the phosphor. Each dot corresponds to a mass position. A saw-tooth voltage obtained from time base A is applied to the vertical input of channel 2 (1A1 preamplifier) in order to displace the successive horizontal sweeps upward. The duration of the saw-tooth is selected to correspond to the total reaction time or some fraction thereof. The A-sweep may be triggered either by the event triggered or by a pre-event trigger pulse. A time delay can be inserted in the external circuit to match the mass positions of the dots with the identical masses observed when channel 1 provides the standard oscilloscope display.

In his earlier version of the instrumentation Lincoln utilized two oscilloscopes: a monitor scope to obtain either the standard oscilloscope display or a z-axis gating film record for a single mass, and a raster scope to obtain the three-dimensional display.[5] The monitor scope also provided the saw-tooth voltage and the amplified mass spectral signals to the raster scope. Using two oscilloscopes permitted the two data recording techniques to be operated simultaneously.

The three-dimensional intensity-modulated raster method provides an extremely useful function. It permits a rapid survey in a single experiment of the time-dependent behavior of many masses (Fig. 6). Specific masses showing evidence of valuable kinetic information can then be selectively monitored by other techniques. Lincoln found that the conjunctive use of this technique with one of the aforementioned synchronized gating methods is simple to implement and is a rapid source of both qualitative and quantitative data.

Two-dimensional linear intensity-modulated display

Kendall devised an intensity-modulated display system which provides a quantitative Polaroid film-record of ion intensity versus time for several masses simultaneously.[6–8] Both mass and ion intensity information are displayed along the vertical axis of an oscilloscope. The horizontal axis is the time axis. Intensity modulation is employed to separate the traces of peak ion intensity versus time for the several masses.

Figure 7 is a block diagram of the instrumental system. The mass spectral signals are supplied to the top vertical deflection plate and a saw-tooth voltage is supplied to the bottom plate of

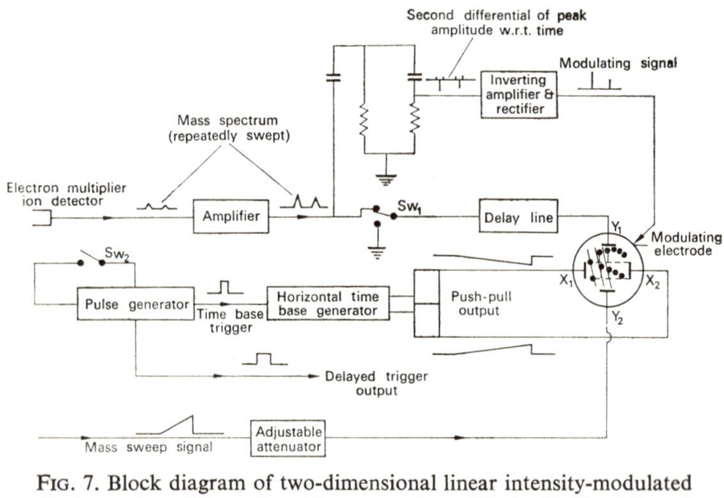

FIG. 7. Block diagram of two-dimensional linear intensity-modulated display system. (Reproduced by permission of B. R. F. Kendall and the American Vacuum Society; ref. 8.)

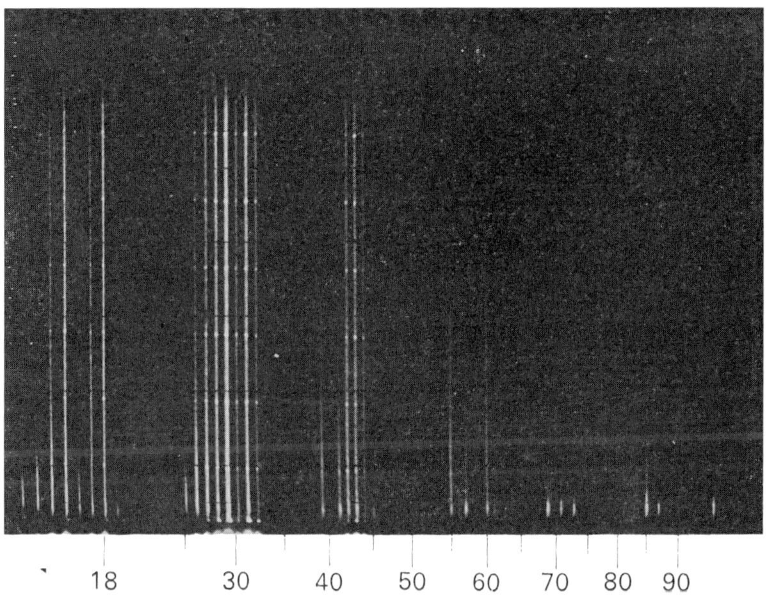

FIG. 6. Three-dimensional raster display for laser-induced vaporization of cellulose. Vertical sweep = 2 msec/cm. Masses 19, 25, 45, 61,73, 82, and 87 are comparatively short-lived, as are masses 58 and 72 which appear at a slightly later time. (Reproduced by permission of K. A. Lincoln; ref. 4.)

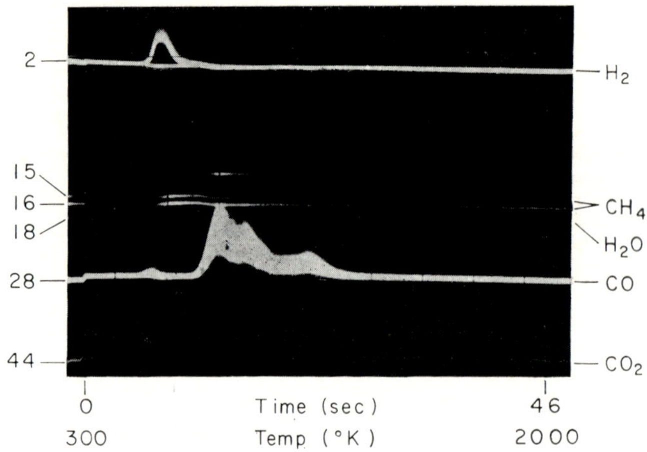

FIG. 8. Two-dimensional display showing changes in gas composition during desorption from a tungsten filament. Adsorption time, 4 hours at residual gas pressure of $1{\cdot}8 \times 10^{-9}$ torr. Largest vertical deflection of CO trace corresponds to 10^{-8} torr partial pressure. (Reproduced by permission of B. R. F. Kendall and the Institute of Physics and the Physical Society; ref. 7.)

a CRT. When the pulse generator initiates a single horizontal sweep at a sweep rate corresponding to many cycles of spectrometer output, the electron beam moves in a down-and-up-and-down zigzag pattern across the horizontal axis. The baseline of the standard oscilloscope display is essentially tipped vertical (slightly less than vertical due to the horizontal component of the beam), while the mass signal deflections from the baseline remain perfectly vertical (due to the narrowness of the signals compared with the total time of a single cycle). A z-axis intensity-modulation is applied to intensify only the peaks of the mass signals. The principle of intensification is the same as for Damoth's z-axis gating method and for Lincoln's three-dimensional raster display. However, Kendall differentiated the mass spectral signals twice and eliminated the negative components to obtain very narrow modulating pulses. When these pulses are applied to an appropriate modulating electrode of the CRT and synchronized with the vertical input by means of the delay line, only the peaks of the mass spectral signals are intensified. By reducing the overall electron beam current, the baseline and the signal slopes can be made invisible, and only the intensified spots where the mass peaks occur are retained.

Absolute mass positions and the zero-amplitude locations for each mass are identified by grounding the top vertical deflection plate and by triggering the mass sweep saw-tooth voltage and the horizontal time base sweep. Under these circumstances, the electron beam is intensified for those masses which exist in the background mass spectrum, but there is no upward vertical deflection to the down-sweeping beam as the beam passes through each mass position. The zero-amplitude mass positions can be based upon interpolation between known masses in the spectrometer background spectrum, can be obtained from a known sample of vapors, or can be recorded from a reproduction of the actual transient experiment. It must be certain, however, that the delay line was correctly adjusted to give synchronization of the modulating pulses with the mass spectral peaks. This can be achieved by increasing the horizontal sweep rate to display only a single cycle of the background mass spectrum and adjusting the delay.

Figure 8 reproduces a film record from one of Kendall's experiments on the desorption of gases from a heated tungsten filament. It shows the partial pressures of gases released during heating from 300°K to 2000°K at 38°C per sec. The markers down the left side of the photograph identify the gases present and indicate the levels from which the corresponding peak heights are to be measured. Kendall stated that the filling-in of the desorption peak contours is due to the limited bandwidth of the circuits used to amplify the modulating signals.[9] He suggested that this can be reduced and made more uniform for all masses by replacing the inverting amplifier and rectifier with a pulse generator. The latter would be triggered by the second-differential signals and would produce modulating signals of constant amplitude. While this improvement would be desirable in Kendall's technique, it could not be applied to Lincoln's three-dimensional intensity-modulated system.

A disadvantage of Kendall's present display system is that an ion intensity versus time profile for a given mass may overlap with the profiles for masses above it. The portion of the mass spectrum that can be displayed on a single oscilloscope may be limited, depending upon the relative magnitudes of the mass peaks and the number of masses that change amplitude in the same time interval.

Multiple trace raster display

The intensity modulation techniques of Lincoln and Kendall incorporate the raster principle of display. However, there is another variation of the raster operation which provides for the standard oscilloscope type of display (i.e. ion intensity versus mass) successively displaced along the vertical dimension of the oscilloscope screen. This method was first recognized by Lincoln[5] and was modified and applied subsequently by Moulton and Michael.[10]

Figure 9 is a block diagram of the raster display system used by Moulton for shock tube chemical kinetics. The significant feature is that gating circuitry for the spectrometer trigger pulses was developed to permit coupling several oscilloscopes in series. The mass spectrum is supplied to one channel of all of the Tektronix C–A dual-input preamplifiers. A ramp (saw-tooth waveform) voltage is supplied to the second channel. The two inputs are then added algebraically, which provides the vertical displacement of the horizontal traces as they are successively triggered. The number of traces per oscilloscope is determined by the number of spectrometer trigger pulses allowed to pass through the gate. The duration of the gate is controlled by a positive rectangular signal supplied from a monostable flip-flop. A supplementary output pulse from the flip-flop is differentiated, and the negative spike generated from the terminating end of the rectangular signal is used to trigger the next stage of ramp and gate pulses. Moulton used five identical circuits and oscilloscopes in series to obtain thirty to thirty-five mass spectral traces per experiment.

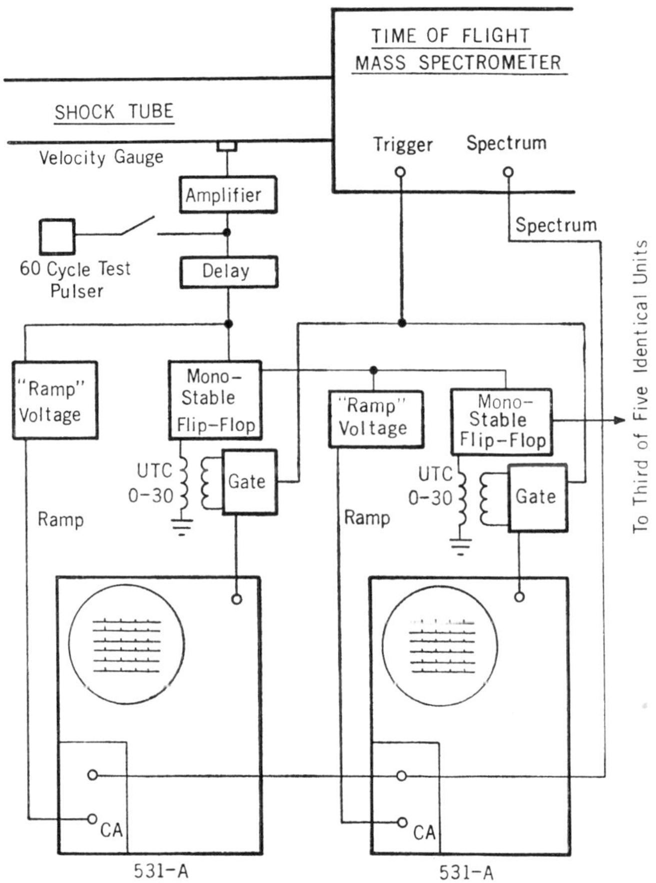

FIG. 9. Block diagram of multiple trace raster display system. The initiating pulse comes from a shock wave velocity gauge. (Reproduced by permission of D. McL. Moulton and the American Institute of Physics; ref. 10.)

The multiple trace raster display was used extensively in studies which coupled the shock tube and the time-of-flight mass spectrometer. An example of the type of film record obtained is shown in Fig. 10. It indicates rapid changes in both neutrals and chemiions during the shock wave induced oxidation of acetylene.[11]

Barton *et al.* also utilized the multiple trace raster display for shock wave studies.[12] However, they have adapted transistorized computer circuitry to count and gate the spectrometer trigger pulses. They report that the circuits operate reliably at counting rates up to 150 kHz.

Two sets of counting registers, which consist of a number of binary circuits, are used. One register contains a number of binaries which is equal to the largest number of mass spectral traces to be displayed on a single oscilloscope. The second register contains binaries equal in number to the number of oscilloscopes to be connected in series. The maximum number of spectral traces that can be recorded is the product of the number of binaries in the two counting registers.

Counting of trigger pulses by the first register is initiated by an event trigger which also activates the first binary in the second register. The latter operation starts the generation of a gating waveform. When a preset number of pulses is recorded by the first register, it switches the first binary of the second register to an "off" position. This operation terminates the gating waveform, resets the first register, and activates the second binary of the second register. This sequence of switching and gating continues. The gating waveforms obtained from the second register are combined with the spectrometer trigger pulses and fed separately to the external trigger inputs of the several oscilloscopes.

Barton *et al.* added three additional features to their raster display system: *z*-axis intensification is applied to the mass spectral signals; a staircase waveform instead of a saw-tooth waveform is used to displace the successive horizontal sweeps on each oscilloscope; a separate oscilloscope raster display provides a superimposed timing-sequence-record of the shock wave velocity gauge signals, spectrometer trigger pulses, and accurate time-interval markers.

A variation of the raster display which reduces the problem of overlapping successive traces was introduced by the Bendix Corporation.[13] A variable horizontal-axis offset permits successive traces to be displayed in the *x*-direction; thus, the signals for a given mass are no longer in vertical alignment. The vertical separation of traces can be reduced, and the number of traces displayed can be increased to a maximum of 64. This is particularly advantageous when only a limited number of masses are present in the spectrum so that interference with another mass position is not a problem. A second new feature in the Bendix system is the provision to divide and superimpose in multiples of 2 a total of 64 successive cycles of spectrometer output among the horizontal raster sweeps of two oscilloscopes. For example, 2 successive cycles may be superimposed on each of 32 successively displaced (vertically and horizontally) traces on one oscilloscope, 4 successive traces may be superimposed on each of 8 traces for two oscilloscopes, etc. A Polaroid record of one variation of the Bendix raster display is given in Fig. 11.

The multiple trace raster technique has three advantages. First, a separate zero-level reference line, common to all masses, is provided for each output cycle of spectral data; this provides for a straightforward measurement of peak heights. Second, with the horizontal axis as the mass coordinate, many masses can be displayed, and interpolation between known mass numbers permits easy identification of unknown masses. Third, the oscilloscope sweep circuits and the external gating circuitry are operable at spectrometer trigger frequencies to above 100 kHz.

Limitations exist for the raster technique, however. The principal one seems to be the small number of spectral traces that can be displayed per oscilloscope for a complex mass spectrum. For oscilloscopes with a 6 cm high display screen, 6–8 traces are about the largest number that can be displayed without overlap and with adequate peak height resolution. Maximum peak height resolution is only about 1 part in 20, due to the finite size of the electron beam spot. A tenfold amplitude range can be scanned with resolution of 1 part in 2 or better. This can be extended by repeating the experiment using increased electron multiplier gain to observe the weaker signals. In addition, several oscilloscopes and cameras are required to cover the full range of rate processes that may be associated with a single transient phenomenon.

As indicated previously, all of the aforementioned data recording techniques utilize Polaroid film recording of an oscilloscope display. It is necessary in most instances to employ a specific set of instrument parameters in order to achieve a good film record that can be quantitatively interpreted. The recommended parameters are: oscilloscope with a CRT potential of 10 kV or greater; P11 phosphor; camera aperture of *f* 1·9 or better; and Polascope type 410 (ASA 10,000) film. This combination should provide adequate contrast for the narrow mass spectral signals or the various representations of them.

Drum camera method

The revolving drum camera recording of a standard oscilloscope display was first employed by Kistiakowsky and Kydd,[14] and was later improved by Meyer.[15] The standard oscilloscope display consists of overlapping traces of mass spectral patterns, which are triggered at the frequency of the mass spectrometer. In the drum camera method, time resolution of these successive traces is obtained by the rapid, vertical motion of a strip of film past the face of the oscilloscope. The resulting film record consists of separate traces vertically displaced on the film (see Figs. 1e and 12). The utilization of three independent coordinate axes to display ion intensity, mass, and time minimizes problems of signal overlapping and mass and time resolution; maximum use is made of the height and width of the oscilloscope display area and of the length of the film strip.

The key to the drum camera method is the combination of the correct set of operating parameters to provide high photographic sensitivity. The following parameters provide the best film record: (1) 24 kV accelerating potential applied to the electron beam of a Tektronix 517A oscilloscope; (2) CRT with a P-16 phosphor (persistence $\simeq$ 0·1 μsec); (3) 35 mm Agfa Isopan Record film developed in Rodinal for 20 min; and (4) Beckman and Whitley model 364-2 oscillographic camera with a *f* 2·3 aperture.[15] The most significant single factor is the 24 kV accelerating potential for the oscilloscope electron beam. Operation at 10 or 12 kV with the other parameters the same does not provide legible film records. *z*-axis brightening provides still greater clarity of the 24 kV records and is desirable to overcome phosphor fatigue. The latter is a decrease in phosphorescence intensity, which progresses from the first to the last of the successive traces.

A photographic contact print of a drum camera film is reproduced in Fig. 12. It shows the decomposition of nitrogen dioxide

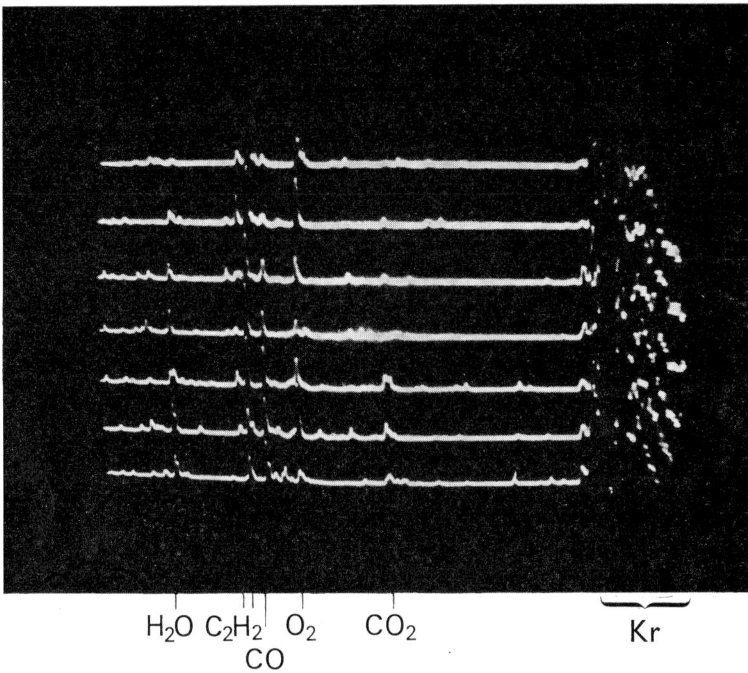

FIG. 10. Raster display of several traces from the shock-wave-induced oxidation of acetylene at 1281°K. The reaction mixture was 96·6% Kr, 1·3% C_2H_2, and 2·1% O_2. Neutrals, fragments, and chemically formed ions are shown. The shock arrived before the beginning of this record. Time runs down, 50 μsec between traces. (Reproduced by permission of D. McL. Moulton; worked performed with J. E. Dove and G. B. Kistiakowsky; ref. 11.)

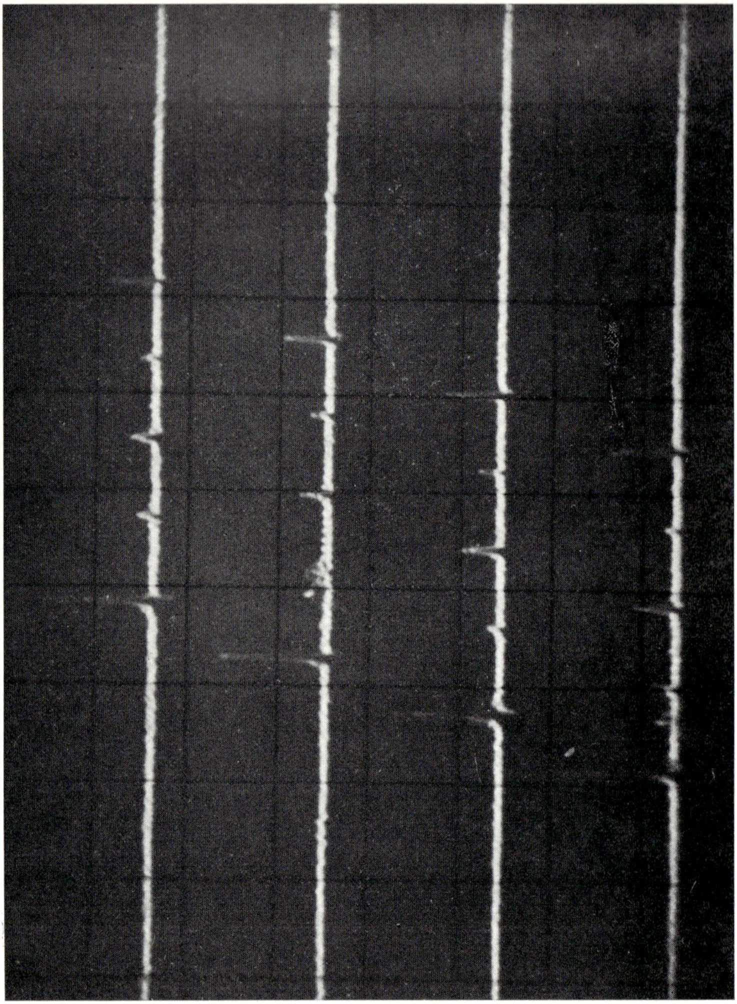

FIG. 11. Mass coordinate (x-axis) offset raster display of spectrometer background gases with 16 superimposed successive mass spectra for each of the 4 traces. (Reproduced by permission of Bendix Corporation.)

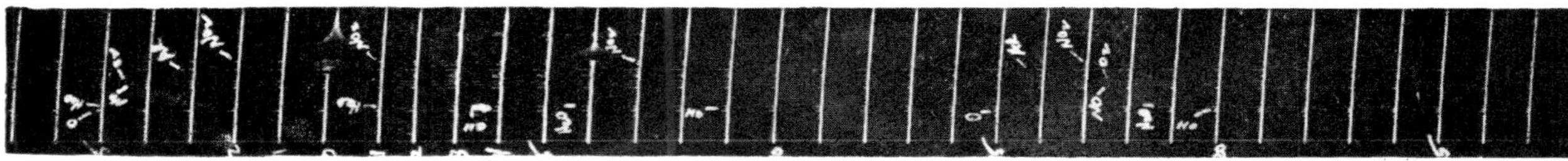

FIG. 12. Revolving-drum-camera record of the flash photolysis of hydrogen and oxygen sensitized by nitrogen dioxide. Reaction mixture was $NO_2 + 3{\cdot}2\ H_2 + 1{\cdot}1\ O_2 + 0{\cdot}5\ Ar$ at 12·0 torr. Film shows decomposition of NO_2 and formation of OH, H_2O, and NO.[16]

and the formation of OH radicals, water and nitric oxide in the flash photolysis of a mixture of nitrogen dioxide, hydrogen and oxygen.[16, 17]

Figure 13 illustrates the auxiliary electronic components which have been assembled to synchronize the flash lamp with the

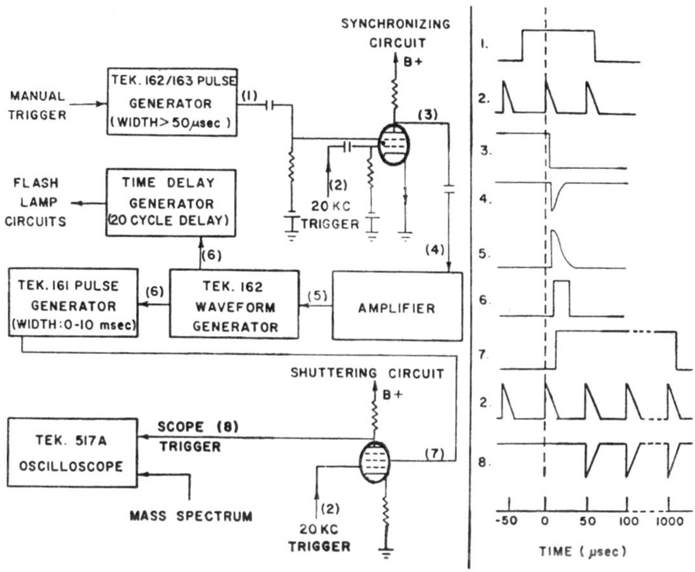

FIG. 13. Electronic components and associated waveforms for synchronization of a flash photolysis and time-of-flight mass spectrometer apparatus. System provides electronic shuttering of the oscilloscope and camera. (Refs. 18, 19.)

20 kHz pulsed operation of the spectrometer and to provide effective shuttering action for the oscillographic camera.[18, 19] The principal component of the synchronizing circuit, which is a modification of a scheme developed by Schellenbaum,[20] is a 5696 tetrode thyratron. The shield grid is biased negatively in the standby position, which prevents indiscriminate triggering of the thyratron by 20 kHz pulses from the mass spectrometer appearing continuously on the control grid. When the desired drum speed of the oscillographic camera is attained, a Tektronix 163 pulse generator is triggered. This produces a positive pulse of duration greater than the 50 μsec interval between spectrometer pulses and of amplitude sufficient to offset the negative bias on the thyratron

control grid. The thyratron fires on the next 20 kHz trigger pulse, and subsequent stages in the circuit become synchronized to the mass spectrometer. The thyratron remains conducting, serving as a lockout against multiple triggering. The output signal is differentiated, amplified, inverted, and applied to a waveform generator which simultaneously triggers a time delay generator and variable-width pulse generator.

The camera shuttering circuit utilizes a 6AK5 pentode. The tube is self-biased at the cathode and for all practical purposes the tube is nonconducting. The peak voltage of the positive 20 kHz spectrometer pulses which are applied continuously to the control grid is insufficient to yield current flow. When the positive gate pulse, which is derived from the variable width pulse generator, is applied to the screen grid the tube conducts and an amplified negative output is obtained. The width of the gating pulse is adjusted to be equal to or less than one revolution of the camera drum. Thus, the shuttering action is achieved by triggering the horizontal sweep generator of the oscilloscope a selected number of times rather than by a mechanical shutter. The additive finite rise times associated with the circuits between the thyratron and the pentode cause the first scope-trigger pulse (Fig. 13, waveform 8) to be one pulse later in time than the 20 kHz pulse which fires the thyratron. Compensation for this 50 μsec lag is achieved with the time delay generator. The principal function of the time delay generator is to delay the firing of the flash lamp for approximately 20 cycles in order to record on film the mass spectral signals prior to the initiation of the reaction.

The complete data recording system has been operated at spectrometer trigger frequencies from 2·5 to 20 kHz and at oscilloscope horizontal sweep rates from 2 to 0·1 μsec/cm. The Tektronix 517A oscilloscope cannot be operated effectively above 20 kHz, although drum cameras are available for faster recording.

The method offers the advantage of recording up to 120 consecutive cycles of spectrometer output with only one oscilloscope and one camera. The peak height resolution is 1 part in 50, and the amplitude range is about thirtyfold. Like the multiple trace raster techniques, the drum camera method records all the masses within a selected mass range, permits rapid, accurate interpolation between known masses, and provides baselines for zero amplitude reference for the actual transient experiment. The latter is particularly desirable in transient experiments for which a.c. or d.c. interferences with the spectrometer output occur. These interferences are readily detected from the film records, and the peak heights can still be measured so long as the signals are not shifted or distorted out of the oscilloscope display.

Other methods

For kinetic processes which are considerably slower than the scanning rate of the mass spectrometer, Marsden *et al.* developed an integrating device which eliminates the statistical fluctuation in peak height.[(21)] The mass peaks in a standard oscilloscope display are individually intensified and submitted to the integrating device. The output of an RC filter can then be fed to a pen recorder, a recording galvanometer, or another oscilloscope. The technique has been applied to mass spectrometer scanning rates of 500 scans/sec, for which the output of each peak is an average of about 50 scans. This system provides much the same function as the analog recorder system of the Bendix spectrometer.[(22)]

Arthur and LePore adapted a quadrupole mass spectrometer to the time resolved observation of two or more mass peaks.[(23)] The system is based upon switching the r.f. and d.c. voltages, which are applied to the quadrupole rods, between two or more preselected values. This results in the alternate focusing of ions of the desired masses upon the electron multiplier. The electron multiplier output is displayed on an oscilloscope. The system provides sampling times variable from 2 msec to 2 sec per mass and requires ion currents of approximately equal magnitude.

PROPOSED DATA RECORDING AND PROCESSING SYSTEMS

While the several oscilloscope display systems already described provide much flexibility and broad applicability as a group, there remains a strong need for a single system with the following characteristics: (1) broad mass range coverage; (2) mass identification; (3) large dynamic amplitude range; (4) high amplitude resolution; (5) high frequency response; (6) high signal-to-noise ratio; (7) capacity to record a large quantity of data; (8) capability to perform successive experiments rapidly; (9) direct conversion of amplitudes and masses to digital information; and (10) computer data processing. In the subsequent paragraphs, efforts are described which may lead to a more versatile data recording system. In particular, the possibilities of magnetic tape recording, analog-to-digital conversion, and computer data processing are considered. Some of these techniques have already been applied to static and very slow mass spectrometric analyses.[24-54]

Magnetic tape recording

Various investigators have considered the idea of recording the output from the spectrometer's electron multiplier on magnetic tape. All masses and all amplitudes for the duration of the experiment would be recorded rapidly and then recovered at a later time. The principal limitation of this technique for time resolved mass spectrometry is the limited frequency response of available tape recorders. The mass spectral signals for the low mass range of a Bendix time-of-flight mass spectrometer are typically 20 nsec wide at half-maximum amplitude. In order to preserve the signal shape and not degrade mass resolution and to maintain amplitude sensitivity, a frequency response in excess of 10 MHz is required. Although many of the oscilloscope systems in use have frequency responses of 20 to 30 MHz, a 100 MHz system is preferable for an advanced system. Tape recorders approaching 5 MHz response are presently available, and it is hoped improved systems will be forthcoming.

Park developed a tape recording system for relatively slow kinetic analyses with a 90° magnetic sector partial pressure analyzer.[55] It has been applied to a study of the reaction of carbon monoxide with oxygen-18 enriched oxygen over a palladium film. The mass range of interest (28–48) was scanned once each second, and the electron multiplier output was recorded with an Ampex CP-100 F-M recorder at 9·5 cm/sec (frequency response: 0–1250 Hz). The multichannel recorder permitted several amplifications of the electron multiplier signals to be recorded simultaneously. A signal proportional to the ion accelerating voltage was recorded on an additional channel to provide mass identification and a real time reference. In order to retrieve information, Park designed a gating circuit for z-axis intensification of one mass at a time from the tape playback to a storage oscilloscope. Playback could thus be accomplished at a speed of 1·52 m/sec. The oscilloscope display was permanently recorded on a Polaroid photograph.

An advancement of this technique would consist of bypassing the storage oscilloscope, except for qualitative analysis, and combining the tape recorder with a digitizer and computer.

Analog-to-digital conversion

One of the most time-consuming tasks in the analysis of time resolved mass spectral data is the translation of the peak heights to digital form for computational use. In the past, this has been accomplished by direct manual measurement with a ruler or magnifier or by a semi-automatic film reading instrument. An example of the latter is the Telereader magnifier, digitizer, and card puncher. The Telereader instrument requires manual alignment of a pair of x- and y-axis crosshairs with both the peak and the base of a mass signal. A "zero" reference level is established on the basepoint, and a digital signal proportional to the peak-to-base separation is obtained from the peak alignment. The peak height and mass coordinate (x-axis) are then recorded on a punched card, one amplitude and mass per card. The assembly of cards is available for direct computer input. Although the amplitude resolution is a few parts per thousand, the film reading is extremely tedious and requires many hours of manual operation.

A considerable improvement in film reading would be achieved with an automatic film scanning and digitizing device. An instrument of this type, called an image digitizer, was developed by Cover at Sandia Laboratories.[56] It consists of a closed circuit television camera, an analog-to-digital unit, and a CDC 160-A computer. Scanning is accomplished in the normal television fashion, left to right and top to bottom. x- and y-coordinates are transmitted to the computer whenever a black to white or white to black transition occurs in the image. With the constant scanning speed, the times at which these points are encountered provide the blueprint for the outline since they are convertible to positions on a rectangular field. By referring to data stored in its memory core, a computer working from a special program could determine the peak heights and mass identifications of all signals in a mass spectral pattern.

Since information is fed into the computer core memory only when boundaries are encountered, storage is more efficient than in the grid method of digitizing, which typically requires the use of one million bits (binary digits) for every image. In the Sandia image digitizer, the number of bits used in the core varies directly with the complexity of the image. The image digitizer could be applied to either Polaroid or 35 mm film records of mass spectral patterns. At present, the resolution of the digitizer is 1 part in 500 in the vertical dimension and 1 part in 660 in the horizontal dimension. Depending upon the number of recorded oscilloscope traces in the TV camera's field of view, peak height resolution would be equal to or less than 1 part in 500. Graphic material in the field of view is completely digitized in a quarter of a second. The principal limitations are in film handling and in computer programming. The former involves rapid, mechanical instrumentation for advancing 35 mm film or inserting Polaroid prints, although this could certainly be accomplished manually with considerable loss in usable digitizer and computer time. The latter involves a complex computer program to identify basepoint and peak height locations on a trace which may have a slanted or distorted baseline. Access to a large computer would probably be required.

Digitization of data recorded in the form of a continuous oscilloscope waveform of ion intensity versus time, e.g. z-axis gating, multiplier anode gating and two-dimensional linear intensity-modulation, can be accomplished by a method developed by Sauer.[57] A device, referred to as the CHLOE system, produces a magnetic tape record of the x- and y-coordinates, at specified intervals, of the various curves on an oscilloscope photograph. The CHLOE requires the data to be on 35 mm film, so the conventional Polaroid photographs must be transformed first. A light beam scans each film frame horizontally and vertically at a scan density of 256 per frame. When the light beam falls on a curve, the CHLOE system registers the coordinates on magnetic tape. A CDC 3600 computer program first separates the array of x-, y-coordinates into the groups corresponding to each curve on the picture and then analyzes the resulting curves for kinetic significance. In the first operation, a testing and curve-fitting operation determines which points belong to which continuous curve and which curve belongs to which baseline. The second operation tests the resulting data sets for linearity in plots of the data according to various kinetic laws.

A vast improvement in operating efficiency and in data accuracy would be obtained if the mass spectrometer output were coupled directly to an analog-to-digital converter (ADC) and a computer-type memory system. Such devices are already available for some slow analysis applications.[50, 53, 54] For very fast time resolved mass spectrometry, however, the principal limitations have been the low frequency response of available ADC units and the cost of such a system. ADC instrumentation with 40 MHz frequency response was recently marketed. In addition, solid-state memory devices with fast switching times and large storage capacity are now available at reasonable cost. It would appear, therefore, that a direct ADC system might be feasible for time resolved data recording.

A system of this type is being explored at Sandia Laboratories. A block diagram of the proposed ADC, memory, and data output instrumentation is shown in Fig. 14. This system would utilize

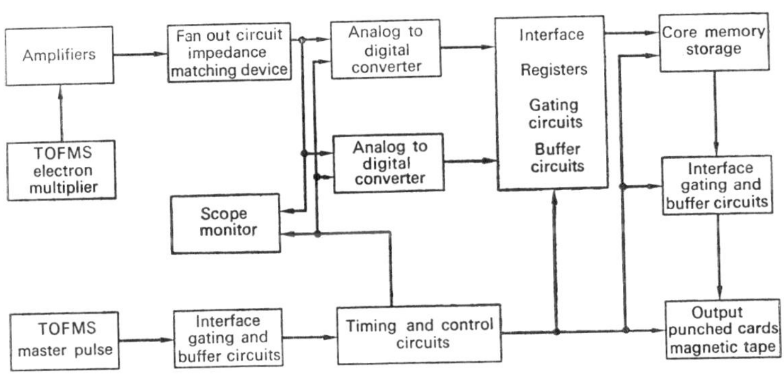

FIG. 14. Block diagram of a possible analog-to-digital converter and computer-memory-storage system for rapid-recording of selected mass ion-intensities.

timing and gating circuits to distribute selected mass signals into separate ADC units. Additional amplifiers (not shown in Fig. 14) would be utilized in the ADC inputs to provide a broad dynamic range and maintain high resolution for the several different mass signals. A single ADC unit for all masses in the spectrum is not yet possible, since the total time required for digitization and storage in the memory devices is still a few microseconds. This time is short enough, however, to permit peak height digitization of signals appearing at a repetition rate of about 200 kHz or slightly higher, which is compatible with the pulsed operation of available time-of-flight mass spectrometers. The memory system would have sufficient capacity to store, perhaps, 1000 data points from each of several successive experiments before transfer to permanent storage on punched cards or magnetic tape would be necessary. The latter would then be used as input for computer processing of the experimental data.

Computer data processing

In analyzing kinetic data obtained by time resolved mass spectrometry, there are in general two stages in the data reduction. The first stage consists of making various corrections and refinements to the raw data so that the final data represent the actual kinetic behavior of the concentrations of the reactant, intermediate, and product species. The second stage consists of subjecting the final data to various kinetic rate law analyses to determine the reaction order, frequency factor, and activation energy. Since the latter subject is treated in many texts on chemical kinetics and since many computer programs have been developed to provide kinetic analyses, it is not discussed here. However, since time resolved mass spectrometry involves some special features, a brief description of a computer program for refining the raw data is provided.

The distinctive feature of time resolved mass spectrometry is the large number ($\sim 10^3$) of data points which can be generated in a single experiment. In addition, time-dependent systematic instrumental variations and impurity signals accompany some transient experiments. A further complication is the changing magnitude of dissociative ionization contributions to mass signals as the parent molecule concentrations change due to the chemical reaction.

A CDC 3600 FORTRAN program, called MASPCTRO, was written to provide data processing incorporating these corrections.[58] Although the program was written specifically for flash photolysis experiments, it is applicable to several kinds of time resolved mass spectrometry research. Figure 15 is a summary flow chart of the calculational program. The program can accommodate 50 masses, 250 equally spaced time-points, and amplitudes ranging from 0 to 9999 (arbitrary units). Ion intensities are referenced point by point in time to an internal, inert standard sub-

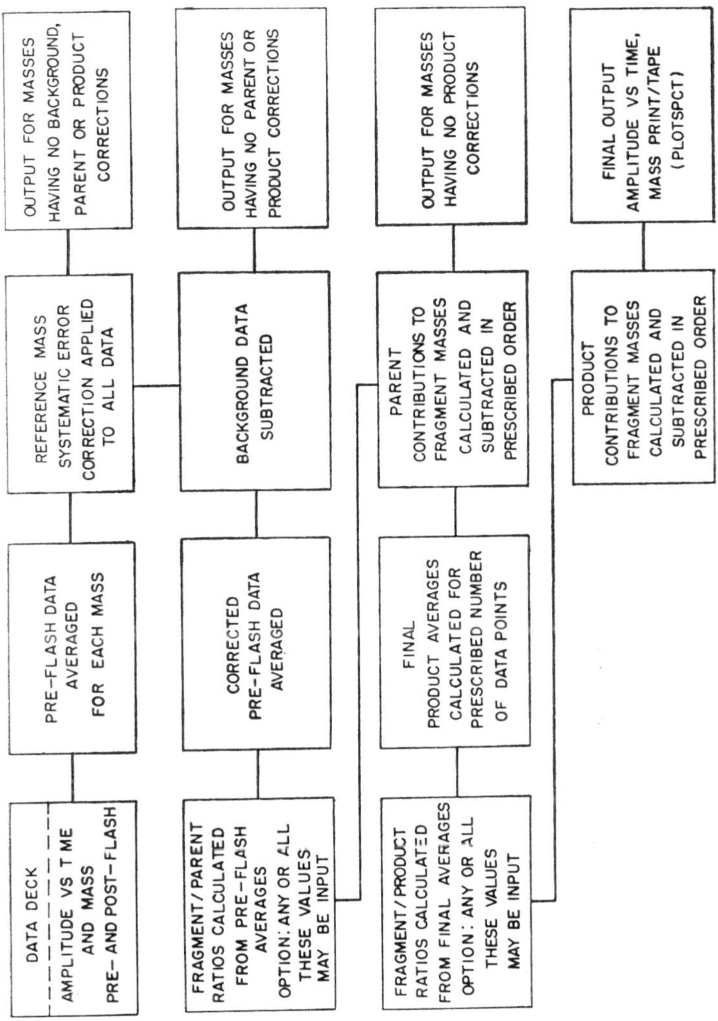

FIG. 15. Simplified flow chart of MASPCTRO computer program for refinement of time resolved mass spectral data. (Ref. 58.)

stance to achieve correction for systematic variations. Background ion and desorbed impurity ion contributions to each mass, obtained from control experiments, are subtracted. Pre-flash data are averaged, and mass-designated ion intensity ratios are calculated to provide fragment-ion to parent-ion ratios. An optional input scheme allows dissociative ionization factors to be supplied when several parent molecules with overlapping mass spectral patterns are present in the reaction mixture. A similar scheme is available for final product analyses. A plotting subroutine permits automatic oscilloscope plotting of ion intensity versus time for each mass. The ion intensity scale range can be fixed or variable. The time scale is absolute, with zero time determined by experimental parameters. The plotting is done by an S-C 4020 high-speed microfilm recorder in an off-line operation. A 9·5 in. camera records the ion intensity versus time data-display for each mass as a 7·5 in. square on sensitized vellum.

Two features are lacking in the present MASPCTRO program. It does not allow for possible spectrometer non-linearity between ion intensity and molecular density, and it does not convert ion intensities to absolute molecular concentrations. The former defect is generally not very significant when the relative concentrations of reacting molecules in the total reactant mixture are small. The latter defect can be easily accommodated by adding another computing stage to the present program or to the subsequent programs utilized for the kinetic analyses.

CONCLUSION

Several simple and useful oscillographic data recording techniques have been developed for use in conjunction with time resolved mass spectrometry. The variety of existing methods allows one to select the one or two techniques which are best suited to a particular kind of experiment. Some experiments may require only a limited amount of qualitative information, while others may require an extensive array of precise quantitative data. It should be expected, however, that the continuing increase in sophistication of scientific experiments will place a greater demand on time resolved mass spectrometers and on the associated data recording and processing instrumentation. The present spectrometers already generate data much faster than the data can be reduced and interpreted. Data reduction and computer processing are separately very costly in both time and money. It is concluded, therefore, that an increased effort is needed in developing integrated systems of mass spectrometers, digitizers, and computers.

REFERENCES

1. DAMOTH, D. C., Paper presented at 8th Annual ASTM Committee E-14 Conference on Mass Spectrometry, Atlantic City, New Jersey, June 1960.
2. DAMOTH, D. C., *Advances in Analytical Chemistry and Instrumentation*, Vol. 4, edited by Charles N. Reilley (Wiley, New York, 1964), pp. 371–410.
3. MEYER, R. T., Paper presented at 15th Annual Conference on Mass Spectrometry and Allied Topics, Denver, Colorado, May 1967; also this book, page 36.
4. (a) LINCOLN, K. A., and WERNER, D., *Ibid.*
   (b) U. S. Naval Radiological Reference Laboratory Report No, USNRDL–TR–68–36.
   (c) LINCOLN, K. A. *J. Mass Spec. and Ion Physics* **2**, 75 (1969).
5. LINCOLN, K. A., *Rev. Sci. Instr.* **35**, 1688 (1964).
6. KENDALL, B. R. F., Paper presented at 10th Annual ASTM Committee E-14 Conference on Mass Spectrometry, New Orleans, Louisiana, June 1962.
7. KENDALL, B. R. F., *J. Sci. Instr.* **39**, 267 (1962).
8. KENDALL, B. R. F., *Trans. Am. Vac. Soc.* (10th National Vacuum Symposium), p. 278 (1963); US Patent 3235725 (1966).
9. KENDALL, B. R. F., Private communication.
10. MOULTON, D. MCL., and MICHAEL, J. V., *Rev. Sci. Instr.* **36**, 226 (1965).
11. DOVE, J. E., and MOULTON, D. MCL., *Proc. Roy. Soc. (London)*, **A283**, 216 (1965).
12. BARTON, S. C., DI VALENTIN, M. A., and DOVE, J. E., *Proceedings 5th Shock Tube Symposium*, Silver Spring, Maryland, April 1965.
13. DAMOTH, D. C., Bendix Corporation, Private communication.
14. KISTIAKOWSKY, G. B., and KYDD, P. H., *J. Am. Chem. Soc.* **79**, 4825 (1957).
15. MEYER, R. T., *Rev. Sci. Instr.* **35**, 1064 (1964).
16. MEYER, R. T., *J. Chem. Phys.* **46**, 967 (1966); also this book, page 31.
17. MEYER, R. T., *J. Sci. Instr.* **44**, 422 (1967).
18. MEYER, R. T., Paper presented at 11th Annual Conference on Mass Spectrometry and Allied Topics, San Francisco, California, May 1963; Sandia Corporation Report No. SC-R-64-185, July 1964.
19. FREESE, J. M., and MEYER, R. T., Sandia Corporation Report No. SC-RR-65-185, May 1965.
20. SCHELLENBAUM, R. L., *Rev. Sci. Instr.* **33**, 1112 (1962).
21. MARSDEN, D. G. H., FORST, W., and FENG, K. K., Rev. Sci. Instr. **36**, 1109 (1965).
22. HARRINGTON, D. B., *Encyclopedia of Spectroscopy*, edited by C. F. Clark (Reinhold Publishing Corp., New York, 1960), pp. 628–47.

23. Arthur, J. R., and LePore, J. J., *Rev. Sci. Instr.* **37**, 794 (1966).
24. Klaver, R. F., Teeter, R. M., Paper presented at 11th Annual Conference on Mass Spectrometry and Allied Topics, San Francisco, California, May 1963.
25. Desiderio, D., and Biemann, K., Paper presented at 12th Annual Conference on Mass Spectrometry and Allied Topics, Montreal, Canada, June 1964.
26. Olsen, R. W., Paper presented at 13th Annual Conference on Mass Spectrometry and Allied Topics, St. Louis, Missouri, May 1965.
27. Olsen, R. W., and Burlingame, A. L., *Ibid.*
28. Cook, H. D., Barker, F. B., and Hudgens, J. E., *Ibid.*
29. McKown, H. S., Cameron, A. E., Christie, W. H., and Trundle, A. S., *Ibid.*
30. McMurray, W. J., and Lipsky, S. R., Paper presented at 14th Annual Conference on Mass Spectrometry and Allied Topics, Dallas, Texas, May 1966.
31. Riley, G. H., and Bacon, D. L., *Ibid.*
32. Stevens, C. M., Walling, D. B., and Moreland, P. E., Jr., *Ibid.*
33. Guibourg, J., Merlivat, J.-C., Bernard, G., Dard, P., and Swiners, B., *Ibid.*
34. Ruth, J. M., Paper presented at 15th Annual Conference on Mass Spectrometry and Allied Topics, Denver, Colorado, May 1967.
35. Hamming, M., Wright, W., Gartside, H., Ford, H., and Haley, J., *Ibid.*
36. McMurray, W. J., Lipsky, S. R., and Green, B. N., *Ibid.*
37. Merritt, C., Jr., Issenberg, P., and Bazinet, M. L., *Ibid.*
38. Boettger, H. G., *Ibid.*
39. Wood, G. M., Jr., and Boies, R. D., *Ibid.*
40. Cone, C., Fennessey, P., Hites, R., Mancuso, N., and Biemann, K., *Ibid.*
41. Eash, M. A., Gohlke, R. S., *Anal. Chem.* **34**, 713 (1962).
42. Thomason, E. M., *Anal. Chem.* **35**, 2165 (1965).
43. McCullough, H., and Krouse, H. R., *Rev. Sci. Instr.* **36**, 1132 (1965).
44. Light, J. E., *Anal. Chem.* **37**, 1627 (1965).
45. Tunnicliff, D. D., and Wadsworth, P. A., *Anal. Chem.* **37**, 1082 (1965).
46. Brauman, J. I., *Anal. Chem.* **38**, 607 (1966).
47. Luenberger, D. G. and Dennis, U. E., *Anal. Chem.* **38**, 715 (1966).
48. Hagan, P. J., and DeLaeter, J. R., *J. Sci. Instr.* **43**, 662 (1966).
49. Howard, H. E., *Anal. Chem.* **38**, 946 (1966).
50. Biemann, K., Bommer, P., Desiderio, D. M., and McMurray, W. J., *Advances in Mass Spectrometry*, Vol. III, edited by W. L. Mead (The Institute of Petroleum, London, 1966), pp. 639–53.
51. Venkataraghavan, R., McLafferty, F. W., and Amy, J. W., *Anal. Chem.* **39**, 178 (1967).
52. Oliver, D., and Simmons, M. C., *J. Am. Oil Chem. Soc.* **48**, 366A (1967).
53. Biemann, K., and Hites, R. A., *Anal. Chem.* **39**, 965 (1967).
54. Burlingame, A. L., Smith, D. H., and Olsen, R. W., *Anal. Chem.* **40**, 13 (1968).
55. Park, R. L., *Rev. Sci. Instr.* **38**, 562 (1967).
56. Cover, R. K., Sandia Corporation Report No. SC-DR-68-812, February 1969.

57. SAUER, M. C., Jr., Argonne National Laboratory Report No. ANL-7113, October 1965.
58. MEYER, R. T., OLSON, C. E., and BERLINT, R. R., Paper presented at 14th Annual Conference on Mass Spectrometry and Allied Topics, Dallas, Texas, May 1966; Sandia Corporation Report No. SC-R-66-928, August 1966.

NOTE ADDED IN PROOF

A three-dimensional mass spectrometer display for gas discharge studies with a quadrupole spectrometer has recently been reported by Engh and Sperling (Paper presented at 16th Annual Conference on Mass Spectrometry and Allied Topics, Pittsburgh, Pennsylvania, May 1968).

DETECTION OF THE NH_2 FREE RADICAL AND OTHER AMMONIACAL COMPLEXES IN THE ELECTRICAL DISCHARGE DECOMPOSITION OF AMMONIA GAS

K. O. Dyson

Department of Physical Chemistry, University of Cambridge

INTRODUCTION

Unstable species are of prime importance in the explanation of the course of many chemical reactions. Their significance in every field of chemistry, from biochemistry through to chemical physics, is patently obvious to workers familiar with these subjects. Until the last 10 years or so the mass spectrometric evidence for unstable species was rather indirect and the information obtained about their properties was qualitative in most cases. The advent of second-generation electronic instruments has contributed to the state where direct evidence for certain species has been obtained and important characteristic parameters have been evaluated.

This article describes studies made with a Bendix 14-101 instrument fitted with an S14-105 ion source. The output equipment was an oscilloscope (CRO) plus an extensive photographic recording system. The mass spectrometer was being used as a general purpose instrument for impurity, end-product analysis, etc., for other research projects. There were periods, however, when the work load was not great and it was felt that the ease of access to the Bendix ion source was such that other reaction arrangements could be coupled to the instrument without impairing its performance in the more utilitarian role. It was decided to use such periods to re-examine early work on the electrical discharge decomposition of ammonia.

The experimental arrangements were dictated by the need

for the whole reactor assembly and supporting vacuum lines to be capable of removal from the instrument so that work of more immediate importance to the department could be performed as and when necessary. The chronological sequence of the experiments was also controlled by this condition. The whole of the research to be described in this paper has been accomplished in "free time", i.e. when there would otherwise have been inactivity in the TOFMS group during the changeover period between other research projects and when time was not being occupied by the more routine tasks.

EXPERIMENTAL

An electrodeless discharge, operating at 2450 MHz and 60 W into a quartz tube, was used in a similar manner to Clyne and Thrush.[1] The coupling device was an air-cooled NBS resonator[2] which could be moved along the quartz tube so that the visible discharge glow could be terminated at any specific point. An inlet jet was inserted between the discharge and the TOFMS sampling leak. The whole reaction tube was co-axial with the fast reaction chamber (f.r.c.) of the ion source and the minimum distance between the discharge point and the ion source leak was 50 cm via 1 cm i.d. pyrex tubing. Certain results obtained with the above reaction assembly indicated the need to reduce the discharge-to-leak distance. It was decided to design a discharge apparatus that could be located adjacent to the sampling leak.

Economic factors dictated the selection of a d.c. electrode reactor rather than the r.f. electrodeless cell which would have been the preferred first choice. The size of the Bendix f.r.c. was an important limiting factor in the design of a suitable system; the final choice is shown in Fig. 1. The reactor consisted of a central stainless steel gas inlet pipe surrounded by a close fitting pyrex tube. This was then surrounded by a second pyrex tube of 1 cm i.d. which extended a few centimetres beyond the end of the gas inlet pipe. Around the inside of the larger pyrex tube an aluminium sheet was sprung to fit closely to the pyrex surface. The distance between the stainless steel pipe (anode) and the aluminium cylinder (cathode) was 1 cm. A connecting wire was run through a drawn pyrex sheath from a metal-through-glass seal to the cathode. The anode connection was made externally to the gas inlet pipe. The whole assembly was then inserted into either a pyrex or stainless steel outer tube which was made a snug fit in the Bendix f.r.c. Vacuum seals were made at the leak wall and the ion source flange as shown on the sketch. A jet was added to the outermost tube so that it coincided with the Bendix f.r.c. gas inlet line; the distance between the cathode end point and the leak was 4·5 cm. The holes in the f.r.c. leak wall and the ion source backing plate were enlarged to 0·055 and 0·2 cm respectively. A 0·0025 cm thick gold foil was inserted against the leak wall. A selection of orifice diameters were available, ranging from 0·0025 to 0·025 cm. The reactor could also be removed from the outer tube without disturbing the latter's vacuum seals and an intermediate flow line inserted as shown in Fig. 2. Inlet jets were added at suitable positions between the reactor and the leak so that preliminary information might be obtained to decide if a kinetic study would be of value.

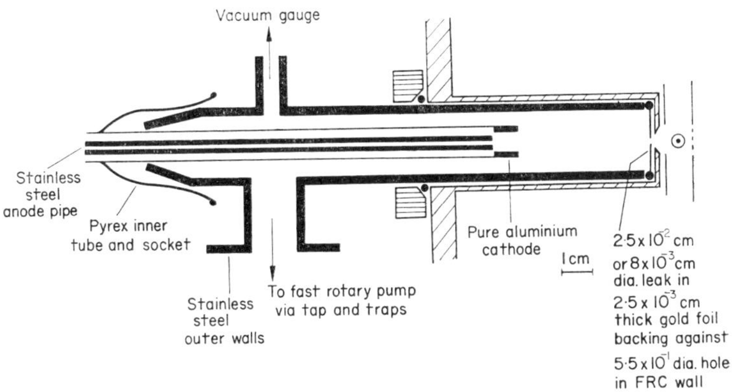

FIG. 1. Coaxial d.c. discharge reactor.

An analogue f.r.c. was constructed from pyrex and the reactor was inserted and operated under varying electrical and gas flow conditions. The same pumping system was used with both the analogue and the Bendix f.r.c. It was found that the reactor behaved in a reproducible manner under controlled, varied conditions. The electrical, pressure and visible glow effects were tabulated and the overall characteristic behaviour of the reactor determined so that any particular visible glow operating mode could be obtained by setting the electrical and pressure parameters. The visible effect of varying these parameters was such that two simple operational modes could be observed. The glow could

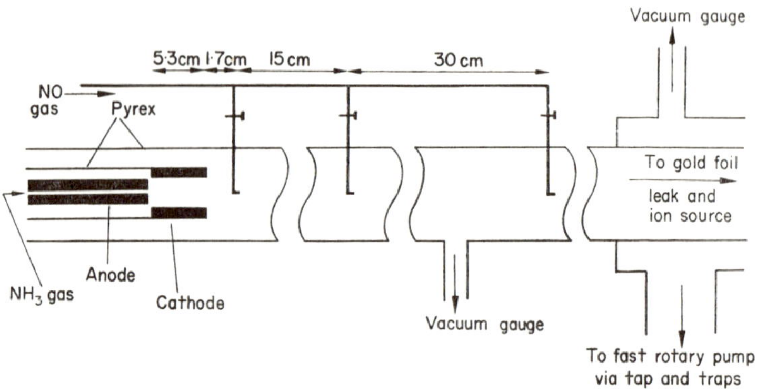

FIG. 2. Intermediate flow line, fitted between the coaxial reactor and the TOFMS sampling leak.

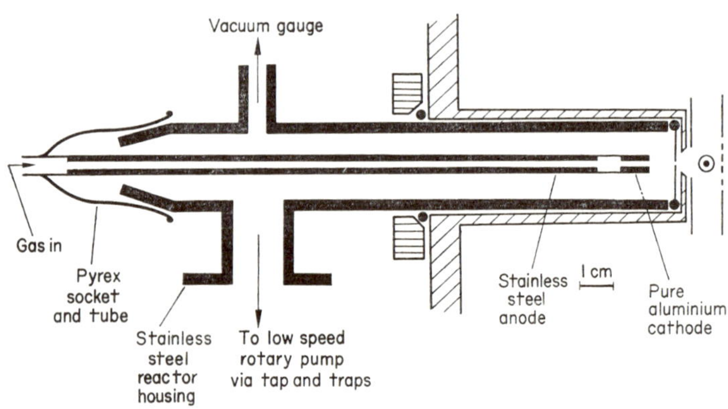

FIG. 3. Small d.c. reactor for use with a slow pumping system. The "discharge ion" experiments were originally performed with this reactor and then repeated, at a later date, with the large coaxial reactor and fast pump combination.

be confined to the region between the electrodes or it could be made to "pop out" through the cathode and stretch downstream. These two modes appeared to be the equivalent states of the "normal" and "abnormal" glows observed in conventional, linear discharge tubes with well spaced, plate electrodes. In all the experiments to be described, with one exception, it is to be understood that the "normal" state of the discharge was used. It was found with this reactor that at all currents up to 10 mA, a "normal" discharge mode was obtained from the minimum operating pressures upwards.

The third type of reactor that has been used is shown in Fig. 3. It is essentially a scaled down version of the preceding design and needs little further description. When a fast rotary pump was not available for pumping the f.r.c. the substitution of this small reactor and the provision of a low-speed pump enabled a study of the ions from electrical discharges in ammonia and other gases to be made. The use of this small reactor, placed near to the leak, together with fast pumping made it possible to observe transient species and to maintain discharges at pressures below the minimum needed by the larger reactor.

The TOFMS was operated in the pulsed mode. All of the measurements obtained before 1966 were made from visual or photographic records of the CRO display. The electron beam pulse width, trap current, frequency of operation, second ion grid and ion lens EHT were all varied in the early work so that the measured ion peaks were in excess of one ion per cycle. The criterion adopted at all times with the CRO detector was that ion signals greater than one ion per cycle would be measured and used for assessment. Lower signal levels were recorded and were of value for qualitative indications, but no measurements were quoted. The highest sensitivity of the basic instrument plus CRO readout was such that a partial pressure of 5×10^{-9} torr argon in the ion source gave an output signal in excess of the required minimum of one ion per cycle.

During 1966 the TOFMS was equipped with a two channel analogue output system and the flight tube was extended from the nominal 100 cm to 195 cm. The availability of the recording output system gave a factor of 10^4 increase in sensitivity and improved presentation of the data. The earlier experiments had provided results which were real and definite, but the information was felt to be qualitative only. The trends and indication observed were useful as pointers to the understanding of the reaction behaviour, but it was decided that the quantitative expressions for concentrations, yields and rates had to await the installation of the electronic recording output system. As soon as the installation was completed the earlier experiments were re-run; the same indications were observed in every case with the addition of quantitative data. A further, very significant, experiment became possible with the increased sensitivity that was available at low electron energies.

Experiments were made to record the emission spectrum of a d.c. discharge through ammonia using the large coaxial reactor. A side-arm was fitted with a converging lens endplate and this was used to focus the emitted glow on to the slit assembly of a Hilger–Watt Glass Wavelength Spectrograph, type D186/7. The whole apparatus was blacked out and exposure times of 40 min were required to produce useable records on Ilford HP3 plates. The electrical, gas pressure and flow parameters were as used in the TOFMS work. The positive column, in a normal d.c. discharge through pure ammonia at low pressures, emits a pale green light overall. Close examination reveals that the cathode edge of each striation has a yellow–red coloured border which is at its most intense in the striation nearest to the cathode. By viewing this striation at an angle it was possible to collect the light from both the border layer and the pale green body. Therefore, a complete record of the whole of the visible emission from the column was obtained. The plates were examined visually and a Joyce–Loebel Recording Micro-densitometer, type E12 Mk. 3, was used to determine the wavelengths of the banded and line spectra by comparison with the known emission standards of a mercury–cadmium lamp source.

The gases used were obtained from BOC, ICI, Matheson cylinders or flasks, or prepared according to established practice. The appropriate vacuum, fractional distillation technique was used in each case to ensure a pure supply. The mass spectrum of the gas was checked and found free from impurities before use in an experiment.

RESULTS

(1) Initial work commenced with the dissociation of ammonia by means of a 2450 MHz electrodeless discharge. With a gas

throughput of 10^{-7} moles sec^{-1} at a flow tube pressure of 2 torr it was found that circa 50% of the ammonia was dissociated to form hydrogen and nitrogen molecules. No atoms or any other products were detected as being of reactor origin 100 cm downstream of the discharge. The 50 eV ratios $NH^+ : NH_3^+$ and $NH_2^+ : NH_3^+$ remained constant whether the discharge was on or off.

(2) The next step was to shorten the discharge-to-leak distance and to increase the gas throughput. With a 50 cm path length, ammonia flow rate of 3×10^{-5} moles sec^{-1} and a flow tube pressure of 0·7 torr it was found that the $NH_2^+ : NH_3^+$ ratio changed from 0·9 : 1 in the off state to 1 : 1 in the on state.

(3) An inlet jet was inserted 10 cm downstream of the discharge glow and mass spectra were recorded with ammonia flowing through the reaction position in the off state and nitric oxide admitted via the inlet jet. When the power was switched on it was found that the NO^+ signal decreased and the H_2O^+ signal increased. Ammonia flow rates 3 to 10×10^{-7} moles sec^{-1}, nitric oxide flow rates 3 to 10×10^{-9} moles sec^{-1}, and flow tube pressures of 1 to 2·5 torr were used; the pattern of results was identical in every experiment. The O_2^+ signal showed no variation with the discharge condition, this indicated that the nitric oxide was not being dissociated by the discharge. The H_2O^+ signal varied with the nitric oxide flow rate.

(4) A coaxial d.c. discharge reactor, similar to that shown in Fig. 1, was constructed from pyrex glass with nickel electrodes. The discharge-to-leak path length was circa 5 cm and the reactor was supplied with up to 8 mA. At ammonia flow rates of 5 to 15×10^{-5} moles sec^{-1} and reactor pressures of 0·2 to 1 torr the 50 eV $NH_2^+ : NH_3^+$ signal ratio changed from 0·9 : 1 to 1·3 : 1 when the power was switched from off to on. At electron energies in the region of 12 to 15 eV the sensitivity of the TOFMS, using CRO detection, was too low for the results to be of value.

(5) A further coaxial reactor, identical to that shown in Fig. 1, was constructed from stainless steel with a stainless steel anode and an aluminium cathode; the discharge-to-leak path length was 4·5 cm. A gold foil leak was used with an orifice of 0·025 cm diameter. The ion signals at m/e 16 and 17 were monitored so that it could be determined if a measurable concentration of NH_2 radicals were being produced, the full procedure that was used being as follows.

Initial scans were made with the reactor at a pressure of 0·05 torr and records were taken of the background spectrum. The ratios of $O_2^+ : O^+$ and $H_2O^+ : OH^+$ were then determined; in the experimental runs the O_2^+ and H_2O^+ signals were monitored and the appropriate corrections applied to the *m/e* 16 and 17 signals. In this way the reactor and ion source background contributions to these signals were effectively removed from the signals due to the ammonia dissociation process. This procedure has been proposed by Lossing *et al.*[3] for use at times when low electron energy experiments are unfavourable. The discharge reactor and associated vacuum lines were then evacuated to 10^{-4} torr for not less than 4 hr prior to the next stage in the experiment. A well-mixed 99% argon : 1% ammonia sample was then admitted to the reactor via a needle valve. When steady flow conditions had been reached and constant pressure readings in the reactor and ion source were observed, the mass spectrum of the mixture was recorded with the reactor power off. The mass spectrum was then recorded with the power on. At pressures above 1 torr no change in the $NH_2^+ : NH_3^+$ ratio was observed for discharge currents up to 10 mA between the off and on states. At lower pressures the off : on ratio increased as the pressure was reduced to a lower limit of 0·2 torr; below this limit it was not possible to maintain a stable discharge. The linear flow velocity of the gas through the reactor-leak volume was circa 700 cm sec^{-1}. Low electron energy runs were made, but were unsuccessful for the reasons quoted previously.

A histogram of a typical result is shown in Fig. 4. Figure 5 shows plots of the variation in the $NH_2^+ : NH_3^+$ ratio for pressures varying between 0·2 and 0·5 torr and for discharge currents of between 3 to 5 mA. The slight spread in the ratio values for the off state is an indication of the inaccuracy inherent in reading peak heights from a CRO photograph. The greater spread in the ratio values for the on state is due to the difficulty in stabilizing the current through the reactor. At low pressures, with the power unit and reactor combination in use at the time, the current fluctuated around a mean level: the limits of the fluctuations have been given above. The significant point was that the $NH_2^+ : NH_3^+$ ratio varied widely for discharge off to on conditions. The effect was reproducible and its magnitude such that no physical variation in the experimental parameters could be responsible except that of the discharge condition.

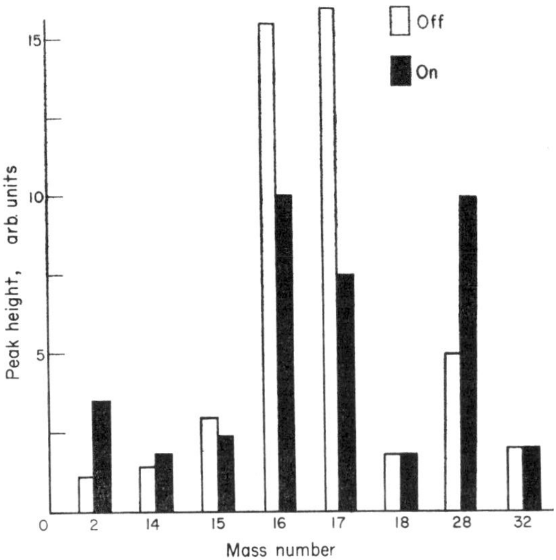

FIG. 4. Histogram plot of peak height versus mass No. for off/on discharge states in NH_3 plus background air.

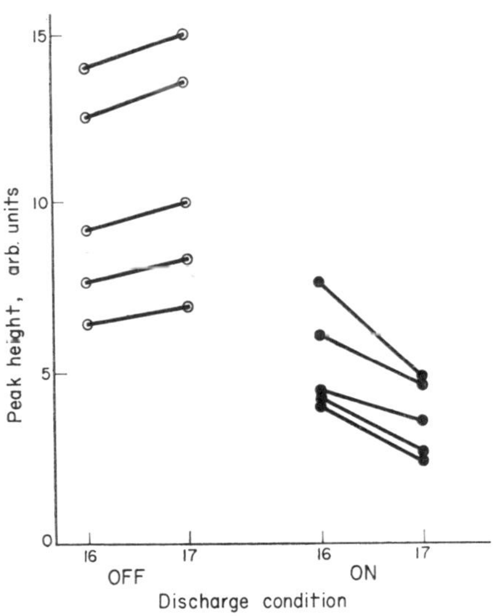

FIG. 5. 50 eV ratio slope plots of m/e 16 and 17 from NH_3, discharge states off/on, for pressures between 0·2 and 0·5 torr.

(6) Experiments were made in which nitric oxide was admitted 1·5 cm downstream of the reactor cathode. Mass spectra were recorded with ammonia flow rates ranging from 3 to 13×10^{-6} moles sec^{-1}, nitric oxide flows ranging from 10^{-9} to 2×10^{-5}

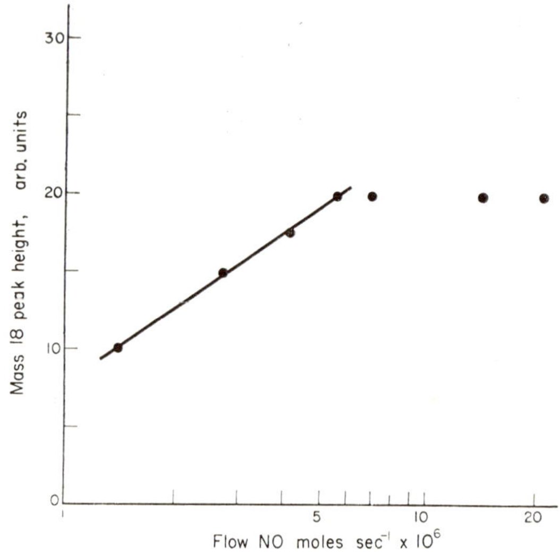

FIG. 6. Plot of ΔH_2O^+/NO flow rate; NH_3 flow rate parameter of $1{\cdot}3 \times 10^{-5}$ moles sec^{-1}.

moles sec^{-1} with the reactor power off and on. A new power unit was constructed with a variable voltage control which enabled a stable discharge current to be maintained to a lower pressure limit. It was found that the change in the H_2O^+ signal from the off to the on state varied linearly with the flow of nitric oxide up to a certain limit. Above this limit of nitric oxide flow no further change in the H_2O^+ signal was observed. Figure 6 shows the plot of ΔH_2O^+ against nitric oxide flow for an ammonia flow rate of $1{\cdot}3 \times 10^{-5}$ moles sec^{-1}. The ΔH_2O^+ signal also varied with the ammonia, but from the few experiments that it was possible to make no definite behaviour could be quoted with confidence.

(7) The coaxial reactor was then moved 75 cm upstream of the TOFMS leak. Three admittance jets were placed in the 2 cm i.d. flow line between the reactor and the leak; the whole assembly was exhausted by a high-speed rotary pump and Fig. 2 shows a sketch of the arrangement. Pure ammonia was passed through

the reactor and nitric oxide could be admitted at any one of the inlet jets. The ammonia was varied from 8 to 26×10^{-6} moles sec^{-1} and nitric oxide flows used ranged from 4 to 65×10^{-8} moles sec^{-1}. The pressure in the flow line was varied from 0·13 to 0·57 torr. The results showed that nitric oxide was consumed and water formed when the discharge was switched from off to on. Plots of $1/[H_2O]$ against time tended towards linearity at the lower pressures reached in the experiments; the plots were definitely non-linear at the higher pressures. The variation of the H_2O^+ signal on the CRO output indicator could be followed as a variation in peak height and also as a variation in peak intensity of the emitted glow from the CRT. The above observation pattern indicated that the $[H_2O]$ was varying from in excess of one ion per cycle to a concentration that was much less than one ion per cycle. Many efforts were made to evolve a system of measurement which would enable data taken from above and below the transition point of one ion per cycle to be treated as a single, pertinent sequence of results. No system was found which stood up to rigorous examination and it was decided that the results obtained from the CRO trace did not allow kinetic data to be extracted; however the overall pattern of the results of this and earlier experiments was consistent and a qualitative explanation of these results was developed.

(8) Experiments were made to obtain the emission spectrum of the ammonia discharge from a coaxial reactor. An ammonia flow of 2×10^{-5} moles sec^{-1} at a pressure of 0·4 torr was pumped through the reactor placed in the analogue f.r.c. mentioned previously. A converging lens endplate, sealed to the "port-hole" of a suitably placed angled side-arm, focused the emission on to the slit assembly of a wavelength spectrograph. The results showed that, within the range of wavelengths covered by the experiment, the emission spectrum of a d.c. discharge in pure ammonia at low pressure consists of three main features:

(a) Nitrogen second positive system bands.

(b) Hydrogen atomic lines.

(c) An extensive, polyatomic band system extending from below 5250 Å to 6460 Å. These limits are arbitrary ones set by the intensity decrease of the bands at the lower wavelengths and by the photographic plate cut-off at the higher wavelengths.

The polyatomic band emission was identified as being due to the α-bands of ammonia,[4–7] the carrier of these bands has been reported to be the NH_2 radical.[8–12] The photographic records obtained with the wavelength spectrograph were examined with a recording micro-densitometer and the band maxima compared with the records reported by Fowler and Badami[7] who used an instrument of higher dispersion. The comparison is shown in Table 1 and the agreement between the two sets of results is considered satisfactory.

TABLE 1. *Emission bands recorded from a d.c. electrical discharge in ammonia at low pressure*

Fowler and Badami band heads	This work band maxima
6652	plate cut-off
6470	6460
6302	6300
6042	6010
5807	5865
5713	5700
5575	5560
5436	5430
5384	5380
5265	5250

Units = 10^{-8} cm. The measurements for this work were made from a spectrograph with a dispersion of circa 90 Å mm^{-1}

Sections 1–8 of these results have been released, in more detail, as separate reports to the research staff at Cambridge. The whole series was issued as a review report.[13] The acquisition of a two-channel output system for the Bendix TOFMS in 1966 provided an opportunity to check the qualitative nature of the earlier results and to permit quantitative re-runs of the more interesting experiments. The increased sensitivity provided by the output equipment also enabled certain new experiments to be performed at low electron energies. All the results reported in (9) and (10) have been obtained using the analogue output equipment and the CRO has been used as an overall monitor. Incidentally, it was regarded as an interesting instrumental exercise to couple the new output equipment to the TOFMS, and to extend the flight tube by 85 cm without recourse to the Bendix service agents. The knowledge that the latter were available bred confidence, but the complete success of these and earlier modifications has demonstrated the ease of adaptation of the TOFMS to any specific research requirement. It is truly a "do-it-yourself" instrument!

(9) Pure ammonia was admitted to the large coaxial reactor which was placed 4·5 cm from the leak. The pressure in the f.r.c. was 0·2 torr and the linear gas velocity in this region was circa 700 cm sec^{-1}. Argon was introduced directly into the ion source by the capillary inlet pipe. The rate of argon admittance was adjusted so that the 30 eV Ar^+ signal was identical, at 10^{-8} A, to that of the 30 eV NH_2^+ from ammonia signal; the ammonia was admitted to the ion source via a 0·025 cm diameter hole in a 0·0025 cm thick gold foil. With the reactor power off the electron energy setting was decreased in steps of 0·25 eV and the NH_2^+/NH_3 and Ar^+/Ar signals were recorded via a stepping switch on to a single channel recorder. The electron energy at which a reading of 10^{-11} A was obtained, a decrease of 10^{-3} from the 30eV value, was accepted as the appearance potential of each ion under the conditions of the experiment in this instrument. The electron energy control was then returned to 30 eV and the original readings obtained. The reactor power was then switched on and the procedure repeated as before. It was noted that the electron energy setting at which the NH_2^+ signal decreased to 10^{-11} A was substantially below the value that was obtained from non-discharged ammonia. The Ar^+/Ar signal was not affected by the discharge and the relevant plots are shown in Fig. 7. A series of experiments in which the reactor state was varied from off to on at each electron energy setting produced identical results.

The electrical resolution of the standard dual-track potentiometer fitted to the TOFMS as the electron energy control is circa 0·1 V. The control shaft had been fitted with an 8 : 1 reduction drive to aid smoothness in operation, but the energy step changes were, arbitrarily, made to be 0·25 eV. These results were not intended to be used to obtain accurate values for ionization and appearance potentials. It is intended to carry out further studies using a modified ionizing arrangement so that this may be accomplished. It was required only to demonstrate that the NH_2^+

signal was present in the ion source at energies below those necessary to obtain NH_2^+ from ammonia. This point was proved and, whilst reservations are held regarding the validity of any attempt to extract more information from these results, it is tempting to compare them with the literature values. The pumping speed of the channel between the tungsten filament and the ionizing region of the ion source was less than 10^{-6} of the pumping speed of the liquid nitrogen trap surface and pumping throat

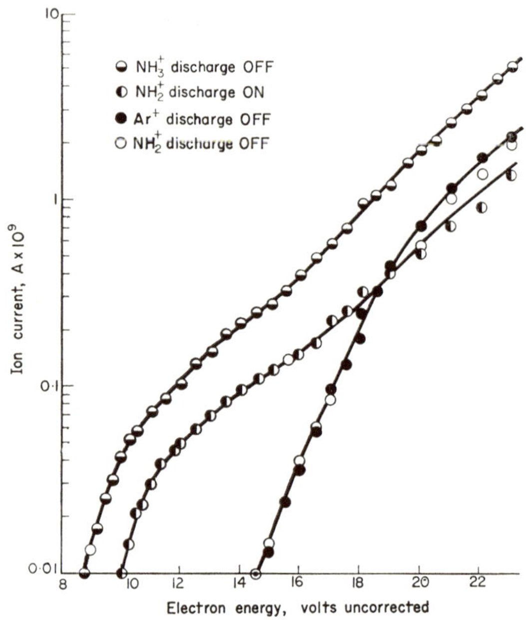

FIG. 7. Plots of ion current versus electron energy for NH_3^+ and NH_2^+ from NH_3 with the discharge off; NH_2^+ from NH_3 with the discharge on; Ar^+ from Ar under both conditions.

exposed to the filament, and interference effects caused by pyrolytic formation of NH_2 radicals on the surface of the tungsten, which then had to travel to the ionizing region, were considered to be negligible in this experiment.

The technique by which these curves were obtained is similar to that used by Kiser and Gallegos[(14)] and Lossing *et al.*[(15)] The electron energy scale may be normalized by shifting the abscissa of Fig. 7 until the Ar^+ signal level of 10^{-11} A is at the accepted ionization potential for Ar^+ of 15·76 eV or 15·75 eV on the Fig. 7

plot. The a.p. for NH_2^+/NH_3 is then seen to be also 15·75 eV and the i.p. for NH_2^+/NH_2 is 4·5 eV lower at 11·25 eV. This procedure is acceptable for the NH_2 radical ion signal since the 30 eV current for NH_2^+/NH_3 with the discharge on was circa 90% of that with the discharge off. The IP of NH_3^+/NH_3 may not be obtained from the plot as the 30 eV current for this ion was 350% of the 30 eV current for NH_2^+/NH_3 with the discharge off. The "simplified procedure" of Lossing *et al.*(15) or Kiser(14) cannot be used in this case. Table 2 compares the values obtained for NH_2^+/NH_3 and NH_2^+/NH_2 with the literature values. The detection of the NH_2 radical in the 1963 experiments, (4) and (5),

TABLE 2. *Electron energy* (eV) *required to form the* NH_2^+ *ion from the neutral molecules* NH_3 *and* NH_2

	Mann *et al.*(16) Exp.	Mann *et al.*(16) Calc.	Lifshitz and Long(17a) Exp.	Lifshitz and Long(17a) Calc.	Melton(18) Exp.	Melton(18) Calc.	This work Exp.
NH_2^+/NH_3	15·7	–	15·8	–	–	–	15·75
NH_2^+/NH_2	–	11·9	–	11·2	11·7	10·5	11·25

by means of the $NH_2^+ : NH_3^+$ ratio change between the discharge off and on states was confirmed by presence of the NH_2^+ ion signal at electron energies between 11·25 and 15·75 eV.

(10) This experiment parallels that of (7) except that the results were obtained by means of the analogue output system with all the consequent advantages that this superior form of readout has for work of this nature. The same result pattern was immediately obvious; at low pressures the plots of $1/\Delta[H_2O]$ against time were linear. At the higher pressures the $[H_2O]$ at J_3 was too low to permit of measurements with the same degree of confidence as at J_1 and J_2 at these pressures; or as at all three jets at the lower pressures. The measurements, made with two ammonia concentrations as parameters, are set out in Table 3. The nitric oxide, with the reactor power on, was increased until no further increase in the H_2O^+ signal was detected and then left set at this level. Therefore, the saturation limit of nitric oxide was being added, the same condition to that indicated by the horizontal portion of the plot in Fig. 6. Test spectra recorded with ammonia flowing through the reactor with the power on, but without any nitric

TABLE 3

Expt. No.	Reactor power state	NO inlet jet	H_2O^+ signal (10^{-9} A)	ΔH_2O^+ signal (10^{-9} A)	$\Delta[H_2O]$ equals $[NH_2]$ radicals (moles cm^{-3})	NO^+ signal (10^{-9} A)
(10a)	off	J_3	0·107	–	–	32·0
	on	J_3	0·245	0·138	$7{\cdot}24\times10^{-12}$	31·5
	on	J_2	0·399	0·292	$1{\cdot}54\times10^{-11}$	20·0
	on	J_1	0·966	0·859	$4{\cdot}50\times10^{-11}$	12·0
(10b)	off	J_3	0·110	–	–	32·0
	on	J_3	0·223	0·113	$5{\cdot}95\times10^{-12}$	31·7
	on	J_2	0·323	0·213	$1{\cdot}12\times10^{-11}$	24·3
	on	J_1	0·480	0·370	$1{\cdot}94\times10^{-11}$	17·4

The other conditions for the experiment were:
(10a) Total pressure 0·284 torr: NH_3 flow $5{\cdot}1\times10^{-5}$ moles sec^{-1}: $[NH_3]$ $1{\cdot}48\times10^{-8}$ moles cm^{-3}; gas velocity 1640 cm sec^{-1}.
(10b) Total pressure 0·140 torr; NH_3 flow $2{\cdot}2\times10^{-5}$ moles sec^{-1}: $[NH_3]$ $6{\cdot}42\times10^{-9}$ moles cm^{-3}; gas velocity 1625 cm sec^{-1}.
The NO flow $5{\cdot}8\times10^{-6}$ moles sec^{-1} and [NO] $1{\cdot}68\times10^{-9}$ moles cm^{-3} were maintained constant throughout the experiment. The figures for $[NH_3]$ and [NO] are for the off/J_3 condition.

oxide flow, showed that the H_2O^+ signal was not affected by the reactor power state. The presence of a trace of oxygen in the ammonia supply can be detected by the increase in the H_2O^+ and the NO^+, N_2O^+, $NO_2{}^+$ ion signals when the power is switched from off to on.

The results shown in Table 3 for the H_2O^+ and NO^+ signals are the arithmetic means of forty-five separate determinations. The measurements were originally taken as percentage deflections of a recorder with a FSD of 4×10^{-9} A. For these "% readings for H_2O" the mean deviation was 4·25%, a standard deviation of 5·44%. It should be stated that this particular experiment was intended only as a preliminary investigation to ascertain the probable value of a full-scale kinetic study. Further work would use a reactor that could maintain a stable discharge at far lower pressures, use a jet assembly that would permit more plotting points along the time axis and investigate other nitrogen–hydrogen molecules in addition to ammonia.

DISCUSSION

A plot of the values for $1/\Delta[H_2O]$ against time, from Table 3, is shown in Fig. 8. In this plot the $\Delta[H_2O]$ has been related to $[NH_2]$ radicals. The equivalence between these two quantities, i.e. the suggestion that the rapid exothermic reaction, reported by Bamford,[24]

$$NH_2+NO \longrightarrow H_2O+N_2,$$

occurs under these experimental conditions is justified by the results previously described. Apart from the well-known accepted

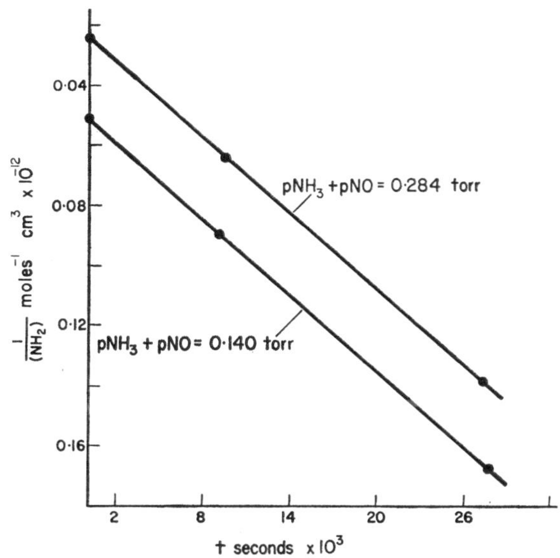

FIG. 8. Reciprocal plot, $1/[NH_2]$ radicals versus time; units of $1/[NH_2]$ moles^{-1} cm$^3\times10^{-12}$ versus seconds$\times10^3$.

reactivity of nitric oxide with other free radical species, it has been shown that;

(a) NH_2 radicals are one of the reactive species present in the products of an electrical discharge through ammonia.

(b) Nitric oxide has been shown to react with one or more of these products.

(c) The concentration of water formed, when nitric oxide is added to the discharged stream of products, varies with the flow rate of nitric oxide up to a certain saturation limit.

(d) Test spectra showed that nitric oxide did not react with non-discharged ammonia under otherwise similar conditions.

(e) There is evidence that nitrogen molecules are present, in greater abundance than is normal from discharged ammonia under these conditions, when nitric oxide is added to the discharged products downstream of the reactor.

The diagnostic peak, therefore, for measurement of the NH_2 radical abundance in the flow tube at various distances (i.e. times) downstream of the reactor is the $\Delta\ H_2O^+$. The $[NH_2]$, in molar terms, can be deduced by the following reasoning. The cross-sections for ionization in a mass spectrometer ion source for water and nitric oxide molecules Q_i are 2·96 and 3·04 respectively. These are figures for ionization by 50 V electrons and are $Q_i \times 10^{16}$ cm^2 quoted by Lampe *et al.*;[19] Rapp and Englander-Golden[20] quote cross-sections in units of πa_0^2 which are numerically different, but the ratio $Q_i H_2O : Q_i NO$ is essentially unity. Therefore we can say that the TOFMS at 50 eV is equisensitive to water and nitric oxide. Accepting that the ΔH_2O^+ signal is diagnostic evidence for the presence of NH_2 radicals and knowing the base [NO] is $1{\cdot}68 \times 10^{-9}$ moles cm^{-3} at an NO^+ signal of 32×10^{-9} A,

$$[NH_2] \text{ radicals} = \frac{\Delta H_2O^+}{NO^+} \times 1{\cdot}68 \times 10^{-9} \text{ moles cm}^{-3}$$

The values for ΔH_2O^+ from column 5 of Table 3 thereby give the values for $[NH_2]$ radicals in column 6 of the same table.

The plot of $1/[NH_2]$ against time is linear so that the rate determining step by which the radicals are destroyed is second order with respect to $[NH_2]$. The detailed reactions which lead to the ultimate destruction of the NH_2 radicals have been variously proposed as[21–36]

$$NH_2 + NH_2 \longrightarrow N_2H_4$$

or $$N_2 + 2\ H_2$$

or $$NH_3 + NH$$

or $$N_2H_4^*$$

and $$NH_2 + H \longrightarrow NH_3^*$$

or $$NH + H_2$$

or a reaction involving a third body as a deactivating intermediary, i.e.,

$$NH_2 + H + M \longrightarrow NH_3,$$

where the third body is ammonia.

The extremely thorough and technically superb survey of the effects of radiolysis by 100 V electrons on ammonia at widely different pressures recently carried out by Melton[18a] seems to have clarified the situation to some extent. The fate of the NH_2 radicals can, reasonably, be held to be due to the following sequence of reactions, on the basis of his results.

$$NH_2 + NH_2 \longrightarrow N_2H_4^*$$
$$N_2H_4^* \longrightarrow N_2H_2 + H_2$$
$$N_2H_2 \longrightarrow N_2 + H_2$$

The work of Melville[23] and Bamford,[24] performed nearly 30 years ago, is not inconsistent with this reaction sequence. Manometric observations of the effects of electrical discharges on nitrogen, hydrogen, ammonia and other gases, made over 30 years ago by Brewer and Westhaver[37] and Westhaver[38] led to their postulating ionic complexes of nitrogen and hydrogen, analogous to the ion cluster mechanisms proposed by Lind,[39] and it may yet be shown that the statement of the complete reaction mechanism explaining the dissociation of the ammonia molecule must include such complexes. Melton's work does not rule out this possibility.

The excellent fit of the $[NH_2]$ values can best be demonstrated by inserting them, individually, into the standard bi-molecular equation for equi-molar concentrations.[40] From

$$\frac{-d[NH_2]}{dt} = k(a-x)^2$$

and integrating to

$$kt = \frac{x}{a(a-x)},$$

we can translate into terms directly compatible with the experimentally observed concentrations. Thus,

$$k_{12} = \frac{1}{t_{12}} \frac{[J_1]-[J_2]}{[J_1]\times[J_2]}$$

where, k_{12} is the rate constant for the decay of NH_2 radicals between J_1 and J_2, t_{12} is the transport time between J_1 and J_2. $[J_1]$ and $[J_2]$ are the $[NH_2]$ radicals at these points. Table 4 summarizes the six different calculations it is possible to make from the two experiments.

TABLE 4. *Experimental rate constants for the decay of* NH_2 *radicals at pressures of* 0·140 *to* 0·284 *torr*

	Expt. (10a)	Expt. (10b)
k_{12}	4·60	4·10
k_{13}	4·26	4·35
k_{23}	4·00	4·75

Arithmetic mean = 4·34.
Units are: 10^9 l. mole^{-1} sec^{-1}.

The transport times between the jets have been taken directly from the expression

$$u = \frac{1}{A}\frac{\varepsilon F}{\varepsilon C}$$

where u is the linear gas velocity, A the cross-sectional area of the flow tube, εF the total gas flow, and εC the total gas concentration.

No correction has been introduced for the mixing process, the reaction

$$NH_2 + NO \longrightarrow H_2O + N_2$$

has been assumed to be completed instantaneously at each jet position. Complete mixing can be expected within 5×10^{-4} sec under the conditions used in these experiments and the maximum error in time determination is circa 5%.

The rate constant for the decay of the NH_2 radicals under these conditions may therefore be quoted as

$$4{\cdot}3 \pm 1 \times 10^9 \text{ l. mole}^{-1} \text{ sec}^{-1},$$

and the large uncertainty has been inserted mainly due to the undoubtedly valid reservations that may be held concerning the acceptance of a more precise value based on three time plotting points. Two previous values have been reported in the literature for this rate constant, one by Hanes and Bair,[28]

$$2{\cdot}33 \pm 0{\cdot}20 \times 10^9 \text{ l. mole}^{-1} \text{ sec}^{-1},$$

and the other by Salzmann and Bair[41] who, after consideration of experiments performed with a TOFMS by Diesen[42] on the thermal decomposition of N_2H_4 in a shock tube, repeated their earlier low pressure (0·425–0·850 torr) experiments at the higher pressures of 5 and 10 torr. They concluded that two separate reactions were responsible for the NH_2 radical decay,

$$NH_2+NH_2 \xrightarrow{k_1} NH_3+NH \tag{1}$$

and

$$NH_2+NH_2+M \xrightarrow{k_2} N_2H_4 \tag{2}$$

where $k_1 = 0{\cdot}46\times10^9$ l. mole^{-1} sec^{-1} and was the extrapolated value, of the observed rate constant, to zero pressure.

$k_2 = 2{\cdot}5\times10^9$ l. mole^{-1} sec^{-1} they state to be the limiting high pressure value for the sum of the two rate constants. Both the experiments in refs. 28 and 41 were monitored by absorption spectrometry, the former using a pulsed r.f. discharge and the latter a multiple flashed photolytic apparatus.

The agreement between the values obtained in the experiments reported in the present paper and the earlier reported values is considered to be satisfactory.

ADDENDA

Some experiments were performed to answer queries posed regarding the possibility of detecting discharge created ions that had survived the sampling process into the ion source. This work was performed prior to the installation of the analogue output system so that absolute concentrations were not determined; however the variety of the ions that may be produced from discharges in various gases is of interest to workers in many fields.

Ions that have been produced in the discharge or during the transit path to the leak (discharge ions) differ in two respects from the normally created ions in the TOFMS ion source (TOFMS ions).

(a) The base width of the discharge ions can be several times that of the TOFMS ions; this is an indication of the large energy spread associated with the use of a discharge as a source of ions.

(b) Discharge ions of a specific m/e ratio occur, on the CRO time base, after that position for a TOFMS ion of the same m/e.

One other important feature that always enables a distinction to be made between the two types of ions is that a different focus pulse value, and often adjustment of the deflection plates and ion lens potentials, is required to obtain the maximum resolution and signal intensity. Therefore there is never any doubt regarding the source of the signals under examination, even if the electron beam in the TOFMS ion source is left on.

(11) A mixture of 1 NH_3 : 10 Ar was pumped through the large coaxial reactor, operated at 10 mA, by a fast rotary pump. The cathode to leak path was 4·5 cm. The ion spectrum was observed with the TOFMS filament switched off, but the instrument was otherwise operated in the normal pulsed mode. A 12 V retarding bias was applied between the reactor cathode and the earthed leak. This was found to give a "cleaner spectrum" with less noise on the CRO trace. The ion *m/e* ratios observed and the variation with pressure are shown in Table 5. The following identities have been assigned:

m/e ratio	*Identity, positive ion*
17	NH_3^+
18	NH_4^+
20	Ar^{++}
32	$N_2H_4^+$
35	$NH_4^+ \cdot NH_3$
40	Ar^+
52	$NH_4^+ \cdot 2\,NH_3$
69	$NH_4^+ \cdot 3\,NH_3$
86	$NH_4^+ \cdot 4\,NH_3$

Apart from the Ar^{++} and Ar^+ ions, all the other ions must be derived from NH_3 and the presence of $NH_4^+ \cdot NH_3$ at the lowest pressures is significant. The result of increasing the gas pressure in the reactor was to cause the higher molecular weight ammoniacal ions to appear, one by one, as the increasing pressure passed certain limits. Decreasing the pressure reversed the effect. NH_4^+ was the most abundant ion at the lower pressures, as the pressure increased NH_4^+ decreased in abundance and $NH_4^+ \cdot NH_3$ increased. The ions of *m/e* 52, 69 and 86 repeated this behaviour as the pressure was increased to the limit where the ion spectrum was no longer evident. The abundances of the latter three ions were very pressure dependent.

A study of ion clusters produced by α-radiolysis of ammonia, over the pressure range 0·05–200 torr, by Hogg and Kebarle[43]

TABLE 5

Reactor pressure (torr)	m/e ratios of ions detected from a normal discharge in an argon/ammonia mixture
0·07–0·15	17, 18, 20, 32, 35, 40
0·15–0·60	17, 18, 20, 32, 35, 40, 52
0·60–1·00	17, 18, 20, 32, 35, 40, 52, 69
1·00–4·00	17, 18, 20, 32, 35, 40, 52, 69, 86
Above 4·00	no ion spectrum observed

The pressures quoted in the table are the limits of the range of pressures over which the different ion spectra were observed. Above 4 torr the discharge reactor was still operating normally, but no ions could be detected in the TOFMS ion source.

has shown the clustered ammonium ion, $NH_4^+ \cdot n\,NH_3$, up to values of $n = 20$. The results reported in the present paper, and in an earlier review,[13] at no times showed a value greater than $n = 4$. Application of Lind's theory of clustering to the ammonium ion would indicate a maximum stability complex at $n = 4$ when the inner ammonium ion would be linked by weak, oscillatory H–N bonds to the four periphial ammonia molecules. The NH_4^+ ion is a perfect tetrahedron in which all the hydrogen atoms are identical and the positive charge is uniformly associated with the whole ion instead of being more on one atom than on another. The ammonia molecule is a pyramid with the nitrogen at the apex and can be cosidered as an incomplete tetrahedron (with N at the centre) with the two unused electrons held on the side away from the three hydrogen atoms. A very simple, pictorial representation of the structure of the $n = 4$ maximum stability complex is shown below; the simplicity of the representation ignores the known inversion oscillation of the ammonia molecule and does not pretend to be a definitive structure. Nevertheless, it does enable an idea of the complex to be held in the mind; a three-dimensional model of the complex would appeal as a more powerful argument. Hogg *et al.*[46] have recently obtained values for the enthalpy changes for the transitions $n = 2, 3$; $n = 3, 4$ and $n = 4, 5$. A discontinuity in the values when $n > 4$ is further evidence that the inner shell (maximum stability complex NH_4^+ 4 NH_3), is completed with four ammonia molecules.

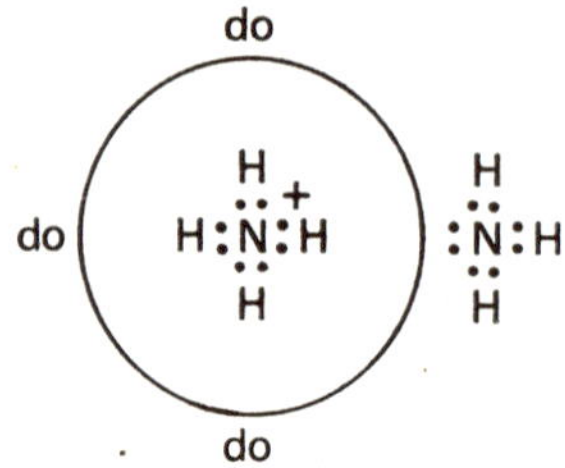

It is possible that the clusters where $n > 4$ in the work of Hogg and Kebarle[43] were caused by the sampling process from the high pressures used in the experiments during which these clusters were observed; the authors recognize this possibility in a footnote to their paper. Nucleation effects are acknowledged to be of importance when using sampling probe jets intermediate to a high pressure and a low pressure region.

Earlier work by Knewstubb and Sugden[44] and Bascombe *et al.*[45] had demonstrated the appearance of ionic hydrated clusters from flames burning at atmospheric pressure and the effect of sampling hole size on the results. In the present paper the orifice diameter was 0·009 cm and the pressure range 0·07–4 torr. The gas flow can be considered as being in the transition region between molecular and viscous flow and nucleation effects caused by the sampling process are not presumed to be of importance. The cluster ion $NH_4^+ \cdot NH_3$ was present at the lowest pressures reached in the experiments where the sample flow through the leak was quasi-molecular. The conclusion then is that the cluster ions observed were formed in the transit path between the reactor cathode and the leak; the marked pressure dependence being a consequence of the probability of cluster formation on the collision frequency. Obviously more work needs to be done on this subject and an experiment in which a beam of NH_4^+ ions could be directed into a controlled pressure of ammonia in a collision chamber situated near to the TOFMS ion accelerating region would be of immense interest.

(12) The same layout was used as in (11), but the d.c. power unit was replaced by a neon sign transformer and a discharge current of 35 mA passed through the reactor. Under these conditions the discharge glow "popped out" through the cathode and extended to within 1·5 cm of the sampling leak. The temperature of the reactor in the previous experiments (1)–(11) had not

been in excess of 320°K; with this current the temperature of the cathode region was circa 360°K. Pure ammonia passed through the reactor at pressures from 0·15 to 3 torr showed the same ions as in (11); the pressure limits were different but the variation pattern was similar. With a mixture of 84 Ar : 1 NH_3 it was possible to maintain a discharge down to below 0·07 torr. At this pressure the ions present were NH_2^+, NH_3^+, NH_4^+ with relative intensities of 1, 10 and 50 respectively. Ar^{++} and A^+ ions were also present in great excess. At 0·5 torr with the same mixture the ions observed were

$$NH_2^+,\ NH_3^+,\ NH_4^+,\ Ar^{++},\ N_2^+,\ N_2H^+,\ N_2H_2^+,\ N_2H_4^+,\ Ar^+$$

The ratio $NH_3^+ : NH_4^+$ was again 1 : 5 and all the other ammonia-derived ions were at 5% NH_4^+ signal or below.

The main points to emerge from (11) and (12) are:

(a) Ammonium cluster ions, $NH_4^+ \cdot nNH_3$, are formed in the transit path between the reactor and the leak. The value for n rises up to a maximum of 4 at 3 to 4 torr for a 4·5 cm path length.

(b) The dominant ions in the discharge spectrum at low pressures are NH_3^+ and NH_4^-.

The two experiments were repeated, with similar results, by using the small co-axial reactor shown in Fig. 3 with a low speed rotary pump. A reactor-to-leak path length of 1 cm was used with a retarding potential of 4·5 V.

CONCLUSIONS

(a) NH_2 radicals have been shown to be present in the products from a low power, normal d.c. glow discharge in ammonia at low pressures. Their existence has been demonstrated by detection at electron energies greater than the i.p. of NH_2^+ but less than the a.p. of NH_2^+ from NH_3.

(b) The NH_2 radical concentration with respect to time has been monitored by a flow tube method and the rate constant for the observed second order decay process has been found to be

$$4{\cdot}3 \pm 1 \times 10^9 \text{ l. mole}^{-1} \text{ sec}^{-1}.$$

(c) The ion output from the discharge at low pressures has been shown to consist mainly of NH_3^+ and NH_4^+ ions. It is precisely at these pressures, under the same conditions, that NH_2 radicals were observed in earlier experiments (cf. reaction mechanisms 6, 7, 8 in the paper by Melton.[18a])

(d) With increasing pressures up to the limit of 4 torr, the ammonium cluster ions, $NH_4^+ \cdot n\, NH_3$, have been shown to be formed with a maximum value for n of 4.

(e) The emission spectrum of the discharge glow emitted from the coaxial reactor has been shown to consist of NH_2 radicals, hydrogen atoms and nitrogen second positive system bands.

ACKNOWLEDGEMENTS

This research project was conceived by Dr. T. M. Sugden when Reader in Physical Chemistry at Cambridge University. His authoritative advice and kindly supervision has been greatly appreciated from the outset. The conclusions that have been drawn from the results, however, are entirely my responsibility. I am grateful to Dr. P. F. Knewstubb for many invaluable discussions and to the assistant staff whose technical support has been of the highest calibre.

Professor R. G. W. Norrish and latterly, Professor J. W. Linnett have been most gracious in allowing the research to proceed as a subsidiary project to the other activities of the TOFMS research group.

REFERENCES

1. Clyne, M., and Thrush, B., *Trans. Faraday Soc.* **58,** 511 (1962).
2. Broida, H., and Chapman, M., *Anal. Chem.* **30,** 2049 (1958).
3. Lossing, F., Ingold, K., and Tickner, A., *Disc. Faraday Soc.* **14,** 34 (1953).
4. Schuster, A., *Rept. Brit. Assoc.* **38,** (1872).
5. Eder, J., *Denkschr. Wien Akad., 60,* 1 (1893).
6. Rimmer, W., *Proc. Roy. Soc.* **103A,** 696 (1923).
7. Fowler, A., and Badami, J., *Proc. Roy. Soc. (London),* **133A,** 325 (1931).
8. Herzberg, G., and Ramsay, D., *Disc. Faraday Soc. 14,* 11 (1953).
9. Walsh, A., *J. Chem. Soc.* **2260** (1953).
10. Swings, P., McKellar, A., and Minkowski, R., *Astrophysics J.* **98,** 142 (1943).
11. Guenebaut, H., *Bull. Soc. France,* 962 (1959).
12. Ramsay, D., *J. Chem. Phys.* **25,** 188 (1956).

13. DYSON, K., *TOFMS Research Report*, July 1964.
14. KISER, R. W., and GALLEGOS, E., *J. Phys. Chem.* **66,** 947 (1962).
15. LOSSING, F., TICKNER, A., and BRYCE, W., *J. Chem. Phys.* **19,** 1254 (1951).
16. MANN, M., HUSTRULID, A., and TATE, J., *Phys. Rev.* **58,** 340 (1940).
17. LIFSHITZ, C., and LONG, F., *J. Chem. Phys.* **41,** 2468 (1964).
18a. MELTON, C., *J. Chem. Phys.* **45,** 4414 (1966).
18b. MELTON, C., and EMMETT, P., *J. Phys. Chem.* **68,** 3318 (1964).
19. LAMPE, F., FIELD, F., and FRANKLIN, J., *J. Am. Chem. Soc.* **79,** 6129 (1957).
20. RAPP, D., and ENGLANDER-GOLDEN, P., *J. Chem. Phys.* **43,** 1464 (1965).
21. MELVILLE, H., *Trans. Faraday Soc.* **28,** 885 (1932).
22. MELVILLE, H., *Proc. Roy. Soc. (London)*, **138A,** 386 (1932).
23. MELVILLE, H., *Proc. Roy. Soc. (London)*, **175A,** 187 (1940).
24. BAMFORD, C., *Trans. Faraday Soc. 35*, 569 (1939).
25. MCDONALD, C., and GUNNING, H., *J. Chem. Phys.* **23,** 532 (1955).
26. MCDONALD, C., KAHN, A., and GUNNING, H., *J. Chem. Phys. 22*, 908 (1954).
27. TAYLOR, H., and BATES, J., *Proc. Nat. Acad. Sci. US*, **12,** 714 (1926).
28. HANES, M., and BAIR, E., *J. Chem. Phys.* **38,** 672 (1963).
29. GEDYE, G., and ALLIBONE, T., *Proc. Roy. Soc. (London)*, **130A,** 346 (1931).
30. BAIR, E., LUND, J., and CROSS, P., *J. Chem. Phys.* **24,** 961 (1956).
31. GEDYE, G., and RIDEAL, E., *J. Chem. Soc.* **151,** 1160 (1932).
32. EMELEUS, H., *Trans. Faraday Soc.* **28,** 89 (1932).
33. TAYLOR, H., and EMELEUS, H., *J. Am. Chem. Soc.* **53,** 562 (1931).
34. BONHOEFFER, K., and FARKAS, L., *Z. Physik. Chem.* **134,** 337 (1928).
35. BONHOEFFER, K., and BOEHM, E., *Z. Physik. Chem.* **119,** 385 (1926).
36. LEWIS, B., *J. Am. Chem. Soc.* **50,** 27 (1928).
37. BREWER, K., and WESTHAVER, J., *J. Phys. Chem.* **34,** 153 (1930).
38. WESTHAVER, J., *J. Phys. Chem.* **37,** 897 (1930).
39. LIND, S., *The Chemical Effects of Alpha Particles and Electrons* (Chemical Catalogue Inc., New York), 1928.
40. MOORE, W., *Physical Chemistry* (Longmans, London, 1956).
41. SALZMANN, J., and BAIR, E., *J. Chem. Phys.* **41,** 3654 (1964).
42. DIESEN, R., *J. Chem. Phys.* **39,** 2121 (1961).
43. HOGG, A., and KEBARLE, P., *J. Chem. Phys.* **43,** 449 (1965).
44. KNEWSTUBB, P., and SUGDEN, T. M., *Proc. Roy. Soc. (London)*, **255A,** 520 (1960).
45. BASCOMBE, K., GREEN, J., and SUGDEN, T. M., *Advances in Mass Spectrometry*, **2,** 66 (1962) (Pergamon Press).
46. HOGG, A., HAYNES, R., and KEBARLE, P., *J. Am. Chem. Soc.* **88,** 28 (1966).

NEGATIVE IONS

J. C. J. THYNNE, K. A. G. MACNEIL and
K. J. CALDWELL
Department of Chemistry, Edinburgh University

1. INTRODUCTION

The formation of negative ions in the gas phase has been known since the early days of mass spectrometry when a number of ions such as H^-, O^- and O_2^- were identified by J. J. Thomson with his mass spectrograph.[1] It might therefore be anticipated that the research carried out on negative ions would be comparable with the extensive investigations made of positive ions and that for both ions the principal features determining their formation and reactivity would be well understood. This is clearly not the case; for example, in a bibliography[2] on mass spectrometry covering the period January 1961 to December 1962, out of a total of nearly 900 papers fewer than 20 were concerned with negative ions.

In general it is the physicist who has been interested in the formation of negative ions, studying such processes as the attachment of low energy electrons to atoms and molecules or the detachment of electrons from negative ions. In this respect much attention has been paid to the oxygen molecule,[3] partly because of the significance of such processes in the ionosphere. The chemist has meanwhile been operating in the very profitable area of positive ions with ion-molecule reaction, appearance potentials, rearrangements and metastable ion studies receiving great attention. At least part of the reason for this relative lack of interest of the chemist in negative ions is in the experimental difficulties associated with the formation and detection of these ions since they are formed much less abundantly than are positive ions. In methane,[4] for example, the most abundant positive and negative ions are CH_4^+ and C^-, and when the ionizing electrons have an energy of 90 eV the ratio of the ion intensities, CH_4^+/C^-, is about 10^4. Also negative ions are frequently formed with excess kinetic energy, a fact which reduces their collection efficiency in the mass spectrometer.

Nevertheless it is clear that the study of negative ions will become of increasing interest and value to the chemist for already their observation in flames[(5)] has suggested that they may play an important part in combustion processes. And in the chemical analysis of organic compounds the use of negative ion mass spectra often offers advantages over that of positive ion spectra for the former are often simpler and more easily interpreted. This is partly because correlation of the mass spectrum with molecular structure is frequently made difficult in positive ion studies by rearrangement reactions which produce ions complicating the overall mass spectrum. In the negative ion mass spectrum there is little indication that, in the initial ionization process, hydrogen migration occurs significantly and consequently the possibilities offered by negative ion studies for linking mass spectra and structure are considerable.

The attachment of low energy electrons to some large organic molecules such as $C_{18}H_{37}OH$ has received a little attention.[(6)] It seems clear that, in such cases, the large negative molecular ion formed does not undergo extensive decomposition and the method suggests itself as being very useful in the analysis of such molecules. Recently Page and his colleagues[(6)] have obtained valuable data regarding the electron affinities of atoms, free radicals and molecules using the magnetron method.

Another illustration of the value of negative ions is in the chemical analysis of electronegative compounds and is provided by the work of Melton[(8)] for a system comprising formic acid HCOOH and formic-d acid, DCOOH. Here the negative ion mass spectra consist of only three components, two of which (O^- and OH^-) are common, the third being $HCOO^-$ and $DCOO^-$. Hence a qualitative analysis of a mixture of these acids can be made by comparing directly the ion currents due to $HCOO^-$ and $DCOO^-$. If the positive ion mass spectrum had been obtained the intensities of twenty-eight positive ions would need to be measured to analyse the mixture.

2. STABILITY OF NEGATIVE IONS

In a neutral atom or positive ion the valence electron moves in an attractive coulombic field modified by the charge distribution of the other electrons at small radii. However, in negative ions there is an extra electron and the force exerted on this electron by the parent neutral is different in character being non-coulombic and of short range. To illustrate this point, the approximate calculations on the hydrogen atom, performed by Massey,[(9)] are reproduced here.

Neglecting polarization effects, the probability that the atomic electron of hydrogen is found at a distance between r and $r+dr$ from the nucleus is $4\pi r^2\psi^2\, dr$, where ψ is the wave function of the ground state of the hydrogen atom. The potential energy $v(r_1)$, due to the atomic electron at a point r_1 from the nucleus, is given by

$$v(r_1) = \frac{4\pi e^2}{r_1}\int_0^{r_1} \psi^2 r^2\, dr + 4\pi e^2 \int_{r_1}^{\infty} \frac{\psi^2 r^2}{r}\, dr,$$

where e is the electronic charge. The first term on the right-hand side arises from the charge within the radius r_1 and the second term from the remainder outside of r_1. The wave function ψ is given by

$$\psi = (\pi a_0^3)^{-1/2} \exp^{-r/a_0}.$$

Here a_0 is written for $h^2/4\pi\mu e^2$, where h is Planck's constant and μ is the reduced mass of the electron–proton system.

Integration of this expression and inclusion of the term $-e^2/r_1$, which arises from the interaction of the extra electron with the hydrogen nucleus, results in the total potential energy at r_1 becoming $V(r_1)$, where

$$V(r_1) = -e^2 \exp^{-2r_1/a_0}\left(\frac{1}{r_1}+\frac{1}{a_0}\right).$$

This gives a field of force decreasing exponentially with distance and not as the inverse square of the distance as does the coulombic field. Including the effects of polarization may alter this force to one falling off as the inverse fifth power of the distance, but this is still decidedly short range.

Now such short-range forces differ from coulombic forces in their properties and, in particular, they give rise to only a finite number of stationary energy states. This result, in conjunction with the Pauli principle, makes it possible to understand the behaviour of different atoms to electron attachment. For example, in the hydrogen atom with one electron, no states are forbidden to the extra electron by the Pauli principle and a stable negative ion can be formed if the effective attractive field is strong enough. In helium, since the one-quantum state is already filled, an extra electron can attach only in a two-quantum state. However, since the effective attractive field is unlikely to be sufficiently strong to give rise to two stationary states, helium is not expected to form a stable negative ion.

It will be roughly true that an atom will form a negative ion if the addition of an extra electron to the neutral results in a closed valence shell. Thus negative ions will be formed by the halogens and also the filling of the ns^2 sub-shells of the alkali metals will result in stable configurations. Similarly the rare gases and the elements in Group II of the Periodic Table are not expected to form negative ions.

3. ELECTRON-MOLECULE ATTACHMENT PROCESSES

When an electron interacts with a molecule, one of several processes may occur depending on the energy of the electron and the nature of the molecule. Of these, the processes which lead to the formation of negative ions are conventionally classified as follows:

(i) Dissociative resonance capture; for a diatomic molecule XY the reaction is

$$XY+e \longrightarrow X+Y^-.$$

(ii) Resonance capture:

$$XY+e \longrightarrow XY^-.$$

(iii) Ion-pair production:

$$XY+e \longrightarrow X^+ +Y^- +e.$$

An elementary and qualitative picture of these mechanisms can be obtained by considering transitions between the potential energy curves representing the ground state of the molecule XY and the molecular ion XY^-. For a diatomic molecule in a stable state the potential energy must be a minimum at the equilibrium separation of the nuclei. When the atoms approach one another, the potential energy rapidly increases due to the repulsion of the charged nuclei and when the atoms separate, the potential energy increases due to the work done against the electronic binding

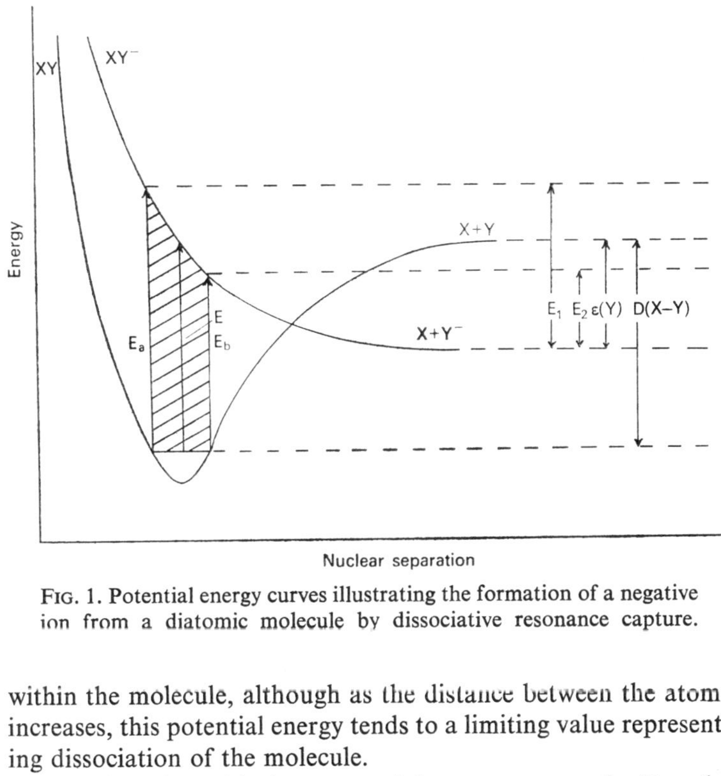

FIG. 1. Potential energy curves illustrating the formation of a negative ion from a diatomic molecule by dissociative resonance capture.

within the molecule, although as the distance between the atoms increases, this potential energy tends to a limiting value representing dissociation of the molecule.

In conjunction with these potential energy curves, the Franck–Condon principle is used. This states that in an electronic transition, because of the large ratio of nuclear to electronic mass, the time taken for the transition is negligible when compared with the nuclear vibration period and hence the nuclear separation does not change significantly, i.e. in the potential energy diagrams, the transition may be represented by a vertical line. In Fig. 1, the shaded region is known as the Franck–Condon region and the most probable transition occurs from the centre of the region requiring an energy E.

XY^- represents the molecular negative ion which, in this case, is formed in an unstable excited state shown by the upper repulsive curve. The ion has an energy greater than at infinite nuclear separation and so breaks up into a radical X and a negative ion Y^-. Electron capture occurs in the limited range of electron energies E_a to E_b and the resulting radical and negative ion possess kinetic energy totalling between E_1 and E_2 and distributed between them, according to the principle of conservation of momentum, as

$$\frac{\mathrm{KE}(X)}{\mathrm{KE}(Y^-)} = \frac{M(Y^-)}{M(X)},$$

where KE represents kinetic energy and $M(X)$ and $M(Y^-)$ are the masses of X and Y^- respectively.

The appearance potential $A(Y^-)$ of the negative ion Y^-, i.e. the energy at which the ion first appears in the mass spectrum, is given by the relation

$$A(Y^-) = D(X-Y) - \varepsilon(Y) + \Sigma W,$$

where $D(X-Y)$ is the dissociation energy of XY, $\varepsilon(Y)$ the electron affinity of atom Y and ΣW is the total kinetic and electronic excitation energies of the species X and Y^-.

This is an example of a dissociative resonance capture process, and in practice usually occurs in the 0–10 eV energy range.

In Fig. 2 some of the possible states of the ion XY^- have an energy less than at infinite nuclear separation. Dissociation occurs only if the electron energy is greater than E and the kinetic energy to be shared between the two species is in the range zero to E_1.

If the electron is captured and the resulting transition has an energy between E_b and E, then a vibrationally excited negative ion $(XY^-)^*$ is formed. This excited ion may release the captured electron and return to its neutral state by the reverse process or it may be stabilized as XY^- by losing its excess energy as radiation or in collision with a third body.

The appearance potential $A(XY^-)$ of the ion XY^- is given by

$$A(XY^-) = \Sigma W - \varepsilon(XY),$$

where $\varepsilon(XY)$ is the electron affinity of XY and ΣW is the excitation energy of XY.

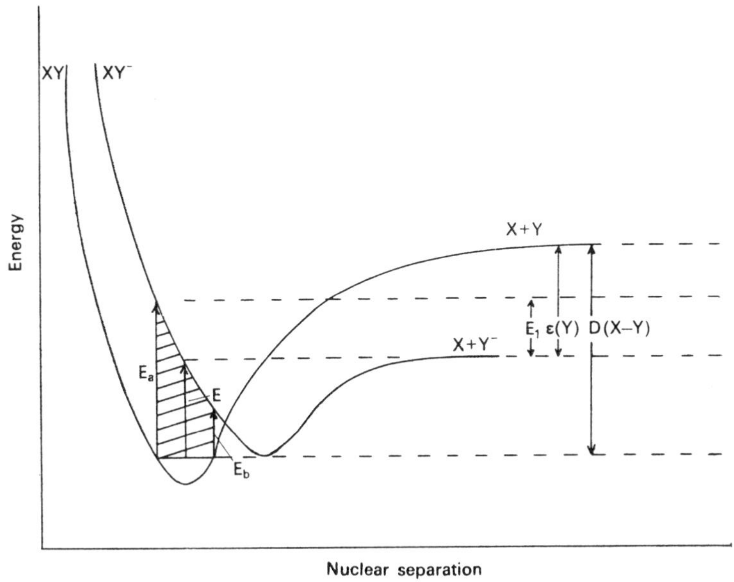

FIG. 2. Potential energy curves illustrating the formation of a negative ion from a diatomic molecule by resonance capture.

Negative ions formed by this resonance capture process usually appear at very low energies ($\sim$ 0 eV).

The third mechanism of negative ion formation is by ion-pair production and is illustrated in Fig. 3.

The molecule XY after suffering electron impact makes an electronic transition into a state which dissociates to give a positive and a negative ion (X^+ and Y^-) either of which may be in the ground state or in an excited state. In this case, the electron is not captured but merely acts as a source of energy and this mechanism can occur over a wide range of electron energies. The kinetic energy to be shared lies in the range E_1 to E_2 and the appearance potential $A(Y^-)$ of the ion Y^- is such that

$$A(Y^-) = A(X^+) = D(X-Y)+I(X)-\varepsilon(Y)+\Sigma W,$$

where $I(X)$ is the ionization potential of the atom X.

Owing to the participation of the $I(X)$ term, the appearance potential of this process usually occurs in the 10–15 eV energy range.

So far only diatomic molecules have been considered but it can be seen that the discussion presented above is applicable at least

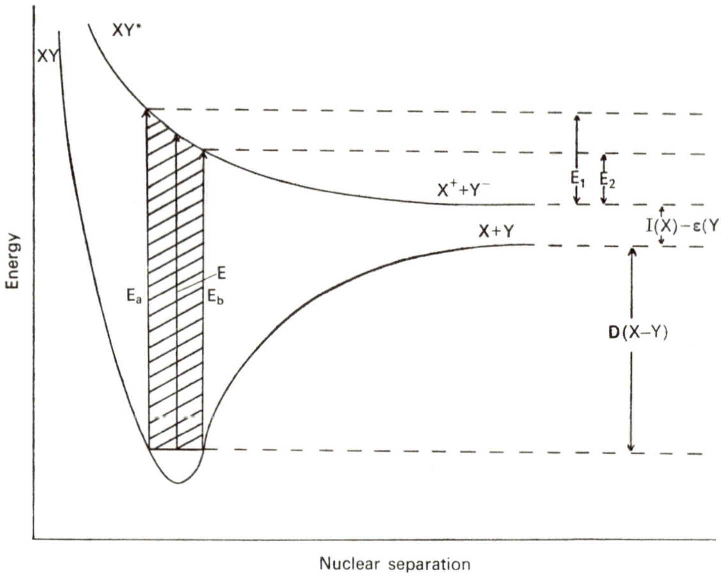

FIG. 3. Potential energy curves illustrating the formation of a negative ion from a diatomic molecule by ion-pair production.

approximately, to polyatomic molecules in which only one bond is broken by the electron impact.

It may also be noted at this point that although in atomic negative ions, the force exerted on the extra electron by the parent neutral is short range and gives rise to only a finite number of stationary energy states, and so makes it improbable that any excited electronic states exist, this is not necessarily the case for molecules. There may be a number of electronic excited states for the molecular ion which do not autoionize rapidly and each of these states has the same assembly of vibrational and rotational levels found in the neutral molecule.

4. PROBLEMS ASSOCIATED WITH ELECTRON-IMPACT STUDIES

In principle, mass spectrometry offers an ideal way of studying negative ions since all of the ions formed in a particular system may be separated and each ionic species identified. In addition the species may be examined over a wide range of energies of the ionizing electrons, this point being of especial importance since, as shown in section 3, different mechanisms for negative ion formation occur in different electron energy regions.

(i) *The filament*

In mass spectrometric work, a heated tungsten filament is frequently used to produce electrons. However, many of the compounds which most readily form negative ions, for example the halogens, carbon tetrachloride, hexafluorobenzene are very reactive and quickly corrode the heated tungsten filament. This has a marked and deleterious effect on the lifetime of the filament and on the energy spread of the emitted electrons.

We have been using a 0·005 in. tungsten filament, the operating current usually being about 2·8 A. When we study negative ion formation in carbon tetrachloride, using a source pressure of about 10^{-6} torr, we find that the average filament has a lifetime of about 8 hr and the time when satisfactory emission is obtained is appreciably shorter than this. As an indication of the effect on the filament of such substances, in Figs. 4–8 we show the formation of the Cl^- ion in CCl_4. This was measured over a small energy range (0–1·2 eV) and the time interval between the first and last figure was about 8 hr. Initially, see Fig. 4, quite a narrow peak indicating a process with a sharp energy dependence is observed at about zero electron energy (the energy scale in the diagrams is uncorrected) with a small second peak being formed about 0·44 eV higher. It is likely that the first peak is a result of the dissociative capture process:

$$CCl_4 + e \longrightarrow Cl^- + CCl_3,$$

and that the second peak is due to the decomposition of an excited state of the CCl_4^- ion. The width of the first peak at half peak height is about 0·14 eV which is a reflection of the energy spread of the electrons emitted from a new filament. Quite quickly, however (Figs. 5–8), the emission from the filament is affected by the presence of carbon tetrachloride, the decrease in height and the broadening of the dissociative capture peak indicating that the energy spread of the electrons has increased. In addition, the second peak increases in size, and as the time in which the carbon tetrachloride is in the presence of the filament increases, so the effects become more marked and the filament becomes unusable. At higher ion-source pressures these degenerative effects become apparent more quickly.

The electron energy scale in Figs. 4–8 has not been calibrated; however it is likely that the maximum of the first Cl^- peak in Fig. 4 occurs near to zero electron volts. If this is so then the onset of the second peak occurs at 0·29 eV and reaches a maximum value at 0·44 eV. This separation of 0·44 eV between the Cl^- peaks may be compared with values of 0·6 eV obtained by Buchel'nikova,[10] 0·7 eV by Fox and Curran[11] and 1·0 eV obtained by Hickam and Berg.[12] Several investigators[13–15] have observed a single broad band rather than two peaks—this possibly reflects the energy spread of their ionizing electrons and the condition of the filament when the measurements were made.

The ratio of the heights of the first and second Cl^- peaks is about 7·5. Fox and Curran[11] reported a ratio of 9·6. Other values observed have been 1·0 by Buchel'nikova[10] and 0·9 by Marriott *et al.*[16] Hickam and Berg[12] obtained a value of 2·0 at room temperature and found that the ratio varied considerably with the temperature of the ion source.

The use of rhenium or thoriated iridium filaments might lead to an improvement in filament lifetime, since the latter particularly has been used in connection with studies of the ionization of halogenated compounds. However, the work of Fox[11] on carbon tetrachloride has indicated that neither of these materials show any considerable improvement over tungsten. However, because of the lower operating temperature required for emission, the use of thoriated iridium does offer the advantage of producing electrons of a narrower energy spread.

(ii) *Regulation of the ionizing-electron beam current*

When examining ionization processes, as well as having accurate knowledge and control of the ionizing electrons [see section 4 (iii)], it is necessary that a constant number of electrons be maintained in the beam or corrections be made for any variations. The number of electrons in the beam is indicated by the electron trap current. In Fig. 9 we show the variation in trap current with electron energy obtained using the Bendix Model 3015 time-of-flight mass spectrometer. It is clear that there is no range of energy over which the trap current is constant. This may be due to focusing problems in the ionizing region.

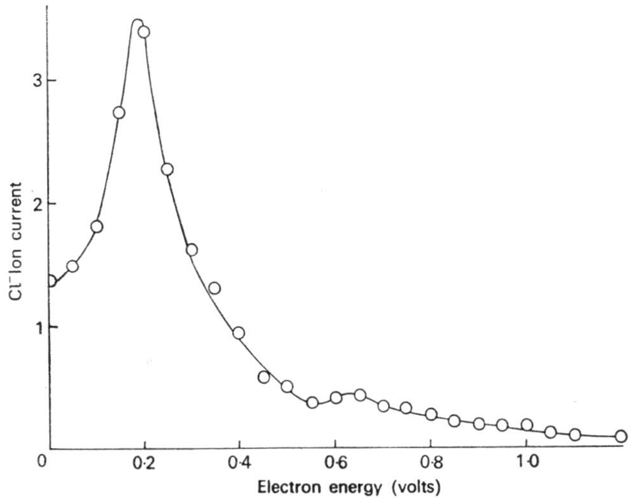

FIG. 4. Ion current due to the Cl^- ion from carbon tetrachloride as a function of the electron accelerating voltage.

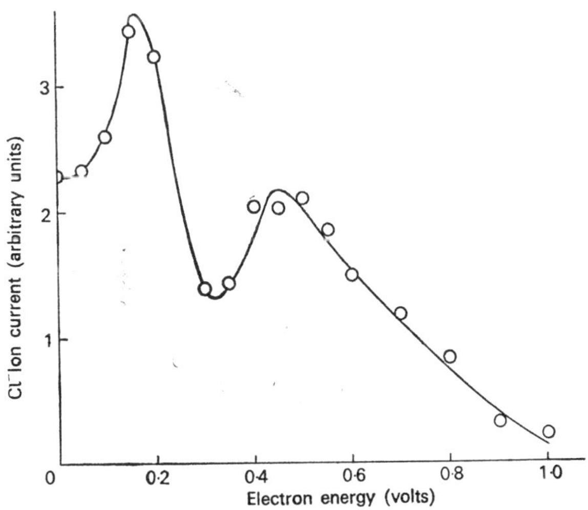

FIG. 5. Ion current due to the Cl^- ion from carbon tetrachloride as a function of the electron accelerating voltage.

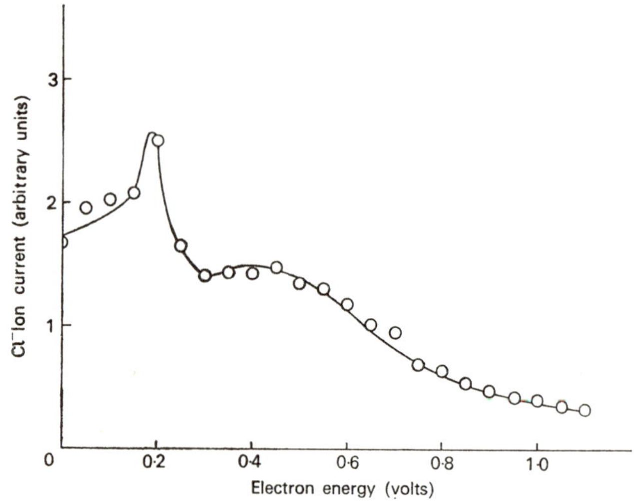

FIG. 6. Ion current due to the Cl^- ion from carbon tetrachloride as a function of the electron accelerating voltage.

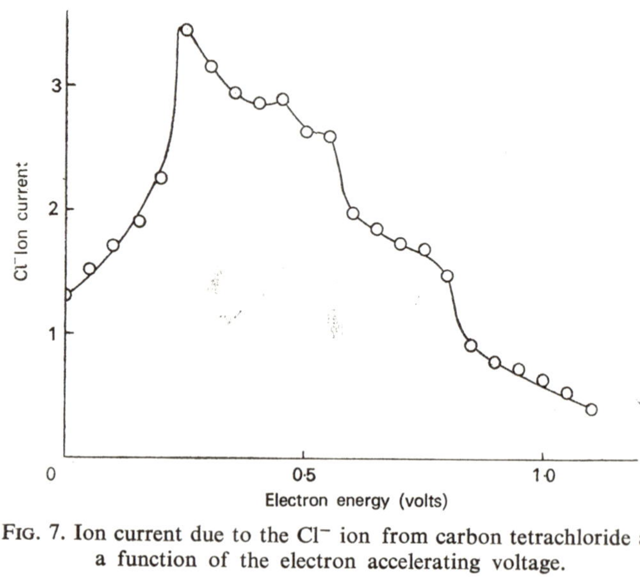

FIG. 7. Ion current due to the Cl^- ion from carbon tetrachloride as a function of the electron accelerating voltage.

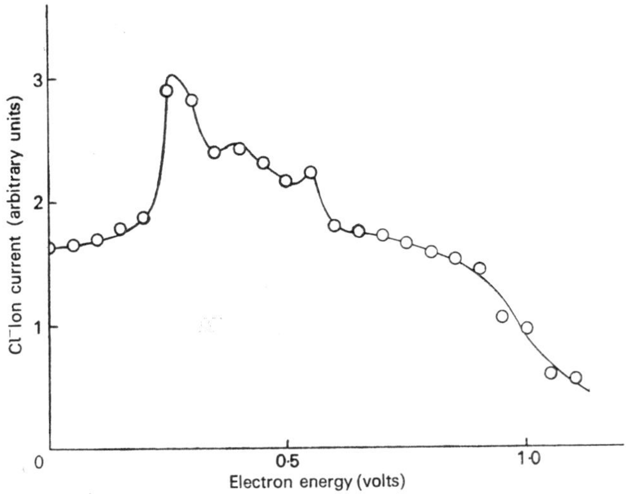

FIG. 8. Ion current due to the Cl^- ion from carbon tetrachloride as a function of the electron accelerating voltage.

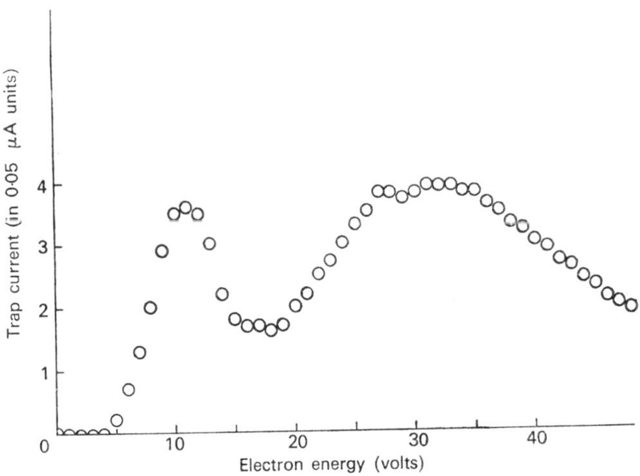

FIG. 9. Electron trap current as a function of electron accelerating voltage in the Bendix Model 3015 mass spectrometer.

It is necessary therefore either to have some method for regulating the trap current or knowledge of the variation of the ion current with the trap current. In the 3015 model, trap current regulation may be achieved either automatically through a feedback control which alters the filament temperature or manually by altering the potential of the control grid in the electron gun. In both the positive and negative ionization modes, when the instrument is operated at electron energies of greater than about

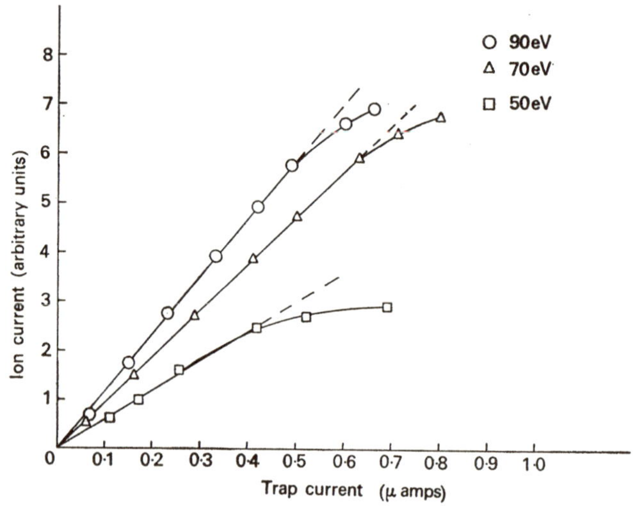

FIG. 10. Ion current as a function of electron trap current at different electron accelerating voltages.

7 or 8 eV, the trap current is large enough to be controlled automatically. Below these energies the trap current is too small to be seen on the microammeter of the instrument and this precludes automatic control. When working at low electron energies, we replace the microammeter by a Vibron electrometer Model 33B-2 and emission control is exercised manually or alternatively the trap current is allowed to vary freely.

Experimentally, it is much more convenient to have the trap current controlled automatically where possible. Both methods of control (automatic and manual) have the disadvantage of altering the energy distribution of the ionizing electrons, but of the two methods, automatic control, which alters the filament temperature, seems to give the more consistent results which suggests that it has the lesser effect on the electron energy distribution.

At low electron energies, where automatic control is not possible, the choice is between manual control and free variation of the trap current followed by the appropriate correction of the measured ion current. In the formation of negative ions the shape and width of the capture peaks which are observed at low electron energies are often almost entirely determined by the energy distribution of the ionizing electrons. Changing the energy distribution by manually altering the potential on the control grid can therefore lead to spurious results being obtained.

Free variation of the trap current followed by correction of the measured ion current is another alternative. In Fig. 10 we show a typical plot of ion current as a function of trap current at three different electron energies. This particular plot was made using a negative ion in the mass spectrum of methanol. The energies shown are high (50 eV and higher) but the behaviour is substantially the same as that observed at low energies.

At each electron energy, the ion current is a linear function of the trap current although in all cases a fall-off from linearity is observed at higher trap currents. The position of the fall-off region is somewhat dependent upon the electron energy. However, if care is taken to ensure that measurements of the ion current are made in the linear region of the plot, correction for the variation in trap current can easily be carried out.

This method has the disadvantage, however, of increasing the uncertainty associated with each experimental point in the ionization efficiency curve since it involves measurement of two variables, i.e. electron current and ion current. It also increases the labour involved in handling any data. Furthermore, we find, especially at low electron energies ($\lesssim$ 7–8 eV), that the trap current tends to increase with time by as much as 100% in the period usually allowed for measuring the ion current at a particular electron energy, i.e. about 2 min.

(iii) *Energy spread of the electron beam*

If useful appearance potential data are to be obtained by the method of electron impact and any fine structure to be seen in ionization efficiency curves, it is necessary that the energy of the ionizing electron beam be homogeneous. Any width in the energy distribution of the beam may cause the point of appearance of an ion to be uncertain and fine structure to remain unresolved. In mass spectrometer studies, the source of the ionizing electrons generally is a heated filament and the emitted electrons consequently have appreciable thermal energies with a spread of 1–2 eV. Clearly, some method is required to select a narrow band of the energy distribution before impact.

Of the various attempts to achieve this, one of the most successful is that devised by Clarke.[17] His 127° electrostatic velocity selector has produced electron beams of only 0·2 eV energy width

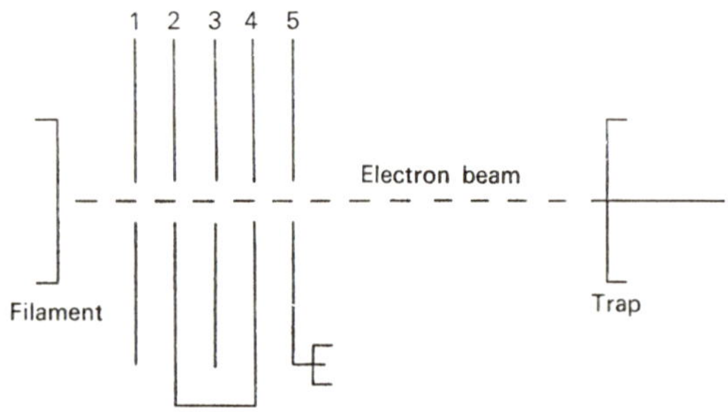

FIG. 11. Schematic diagram of the arrangement of electrodes in the Fox-type electron gun.

and this has resolved fine structure in the ionization efficiency curve of the ion N_2^+ from N_2. Improvements have been made by Marmet and Kerwin[18, 19] reducing the energy spread to 0·04 eV and increasing the beam intensity, a factor which also is of great practical importance.

Perhaps the most widely applied method is that due to Fox *et al.*[20] called the retarding potential difference method (r.p.d.). In the Fox modification of the conventional electron gun, an extra electrode or grid is introduced between the filament and the ionizing chamber (Fig. 11).

The Fox, or retarding electrode, 3, is biased slightly negative with respect to the filament by an amount Vr. This removes from the beam the component of energy less than Vr and results in a sharp low energy cut-off in the beam. By increasing the negative potential of this electrode by a small amount, ΔVr, a further energy slice is removed. The resulting change in ion current, Δi, which occurs when ΔVr is applied, is then due to a band of electrons homogeneous to within ΔVr. The accelerating potential applied to the beam V is the potential difference between electrode 3 at its most negative, i.e. where ΔVr is applied, and the ionization chamber. Repeating the process at different values of V allows an ionization efficiency curve to be constructed by plotting Δi as a function of V.

Electrodes 2 and 4 can also be included (called the five-electrode method) and held at a potential close to that of the retarding electrode. These create an equipotential region around electrode 3 and provide a sharper energy cut-off.

The system is also pulsed; electrode 1 being biased positive with respect to the filament in the first part of the operation cycle allowing electrons to pass into the ionization chamber. At the same time, the ion draw-out voltages are off. Any broadening of the electron energy distribution due to the presence of an unknown electric field in the ionization chamber is thus reduced. In the second part of the cycle, electrode 1 is biased negative with respect to the filament, cutting off the electron beam, and the ion draw-out voltages are switched on.

Melton and Hamill[(21)] have given details of the modifications required in converting the Bendix Model 12–107 source for use in the r.p.d. mode. Using this source they have observed fine structure in the ionization efficiency curves of some positive ions including $C_2H_6^+$, $C_2H_5^+$, $C_2H_4^+$ from ethane and Kr^+ and A^+. They have also[(22)] examined Xe^+, CH_4^+, CD_4^+ and parent and parent-minus-one-hydrogen ions from C_2H_6, C_3H_8 and C_4H_{10}.

Some of the negative ions which have been studied by the r.p.d. method are shown in Table 1.

The Bendix Model 3015 time-of-flight mass spectrometer, which was used in the experimental work described in this review, is equipped with a Fox-type source consisting of five electrodes. This enables the three-electrode or five-electrode method to be used. The five-electrode method has been tried in conjunction with SF_6 in order to examine the resulting spread in the energy of the ionizing electrons. SF_6 forms a negative ion, SF_6^-, with great facility at low energies by a resonance capture process, and it is generally accepted that this occurs over a very narrow energy range compared with usual electron energy distributions. The measured width of the capture peak is then almost entirely

TABLE 1. *Negative ions studied by the r.p.d. method*

Molecule	Ions studied	Reference
H_2	H^-	(23)
Cl_2	Cl^-	(24)
Br_2	Br^-	(24)
I_2	I^-	(24) (25) (26) (27)
HF	F^-	(28)
HCl	Cl^-	(28) (29)
HBr	Br^-	(28)
HI	I^-	(28)
NO	O^-	(30)
O_3	O_2^-, O^-	(31)
SO_2	SO^-, O^-	(32)
CO_2	O^-	(32)
CS_2	CS^-, S^-, C^-	(32)
H_2S	HS^-, S^-	(32)
NO_2	O^-	(33)
N_2O	O^-	(34)
PCl_3	PCl_2^-, Cl^-	(35) (36)
$SbCl_3$	Cl^-	(37)
NH_3	NH_2^-	(32)
CCl_4	Cl^-	(38)
CCl_3F	CCl_3^-, Cl^-, F^-	(39)
$POCl_3$	PCl_2^-, Cl^-	(35) (36)
PCl_5	Cl^-	(36)
SF_6	SF_6^-	(40)
	SF_6^-, SF_5^-	(41)
	F^-	(42)
	SF_6^-	(43)
$W(CO)_6$	$W(CO)_5^-$, $W(CO)_4^-$, $W(CO)_3^-$, $W(CO)_2^-$	(44)
$Cr(CO)_6$	$Cr(CO)_5^-$, $Cr(CO)_4^-$, $Cr(CO)_3^-$, $Cr(CO)_2^-$, $Cr(CO)^-$, Cr^-	(44)
$Mo(CO)_6$	$Mo(CO)_5^-$, $Mo(CO)_4^-$, $Mo(CO)_3^-$, $Mo(CO)_2^-$	(44)
$Fe(CO)_5$	$Fe(CO)_4^-$, $Fe(CO)_3^-$, $Fe(CO)_2^-$, $Fe(CO)^-$	(44)
C_7F_{14}	$C_7F_{14}^-$	(43)

due to the energy distribution being used, the measured shape of the capture peak being the electron distribution reversed on the energy scale since clearly the higher energy electrons in the distribution reach the critical energy for capture before those of a lower energy.

Figure 12 shows a measured capture peak obtained by biasing electrode 3 by $-0{\cdot}2$ V with respect to the filament, electrodes 2 and 4 by $-0{\cdot}2$ V and chopping electrode 3 a further $-0{\cdot}2$ V. "Negative" voltages were obtained by introducing a 3 V dry cell into the electron energy circuit. The figure shows a smooth energy distribution with a width of about 0·3 eV at half-peak height. However, the shape of the distribution is not ideal for

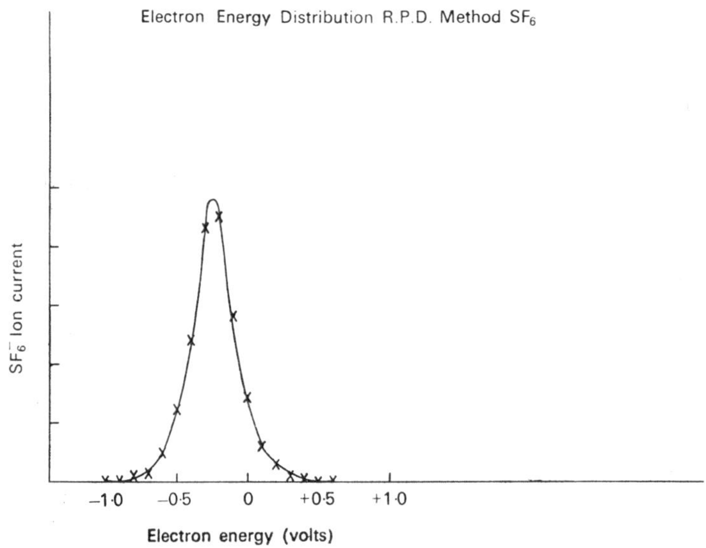

FIG. 12. Ion current due to the SF_6^- ion from sulphur hexafluoride as a function of the electron accelerating voltage obtained using the r.p.d. method.

experimental use since it contains a considerable high energy tail shown by the long low energy onset of the capture peak. With the present instrument, this tail has not yet been successfully eliminated.

One of practical difficulties of the r.p.d. method is that it involves measuring the difference between two ion currents, this difference depending on the size of the chop voltage ΔVr, applied to the retarding grid. The smaller the chop voltage, the more homogeneous the "slice" of electrons selected but in order to obtain a measurable Δi, especially with low intensity negative ions, it has been found that a chop voltage of $\sim -0{\cdot}2$ V is close to the smallest value that is practicable.

The r.p.d. method has been criticized by Marmet[45] from arguments based on the experimental work of Hartwig and Ulmer.[46] Marmet has pointed out that once past the retarding electrode, the sharp energy cut-off is not maintained but the beam relaxes back to a symmetrical distribution. From this he has shown that, from the r.p.d. method, "ionization efficiency curves" may in fact be compounded of the true ionization efficiency pattern plus some unknown fraction of its derivative. However, the fact that in several cases breaks found in positive ion ionization efficiency curves coincide with known excited electronic levels of the ion cannot be disregarded.

What does seem likely is that the electron beam does not maintain its sharp energy cut-off caused by the retarding electrode but that a relaxation process results in a new distribution. Consequently the difference in ion currents measured by r.p.d. is not due to an almost monoenergetic electron slice but to the difference between two slightly displaced energy distributions of unequal intensity.

This idea is supported by the recent work of Winters *et al.*[47] who have produced a third method, called the energy distribution difference (e.d.d.) method, of removing the effect of a broad energy spread. They have treated the problem in the following way: if the electron energy distribution normally produced is Maxwellian or at least contains an exponentially decreasing high-energy tail, the effective width of this distribution may be reduced, by subtracting from it, a fixed fraction of the distribution at a slightly higher accelerating energy. Thus if the electrons possess a distribution of kinetic energy U, such that the probability of finding an electron in the range U to $U+dU$ in unit time is $f(U)$ and they are accelerated by a potential V, then the total electron energy E is given by

$$E = U+V \quad \text{and} \quad f(U) = f(E-V).$$

At a slightly higher accelerating potential V', the distribution is described by $f(E-V')$. Multiplying the second distribution by some factor b and subtracting this from the first distribution gives a new, narrower distribution $\Delta f(E-V)$ such that

$$\Delta f(E-V) = f(E-V) - bf(E-V').$$

The constant b is chosen to eliminate most of the high energy tail and to make the new distribution as narrow as possible. For example, with a filament temperature of 1730°K the resultant Maxwellian distribution has a width at half-height of about 0·5 eV. By using $V'-V = 0{\cdot}1$ eV and $b = 0{\cdot}63$, the width at half-height of the difference distribution is only 0·14 eV.

The ion current $i(V)$ is measured at small electron energy increments of ΔV and the ionization efficiency curve constructed by plotting $[i(V)-bi(V+\Delta V)]$ as a function of V.

This method has resolved fine structure in the ionization efficiency curves of the ions A^+, Kr^+, Xe^+, N_2^+ and O_2^+. It has not been applied to any negative ion studies to date.

Finally, analytical methods have been used[48] to remove the effect of the electron energy distribution from ionization efficiency data. These techniques involve smoothing of experimental data and deconvolution using iterative processes or Fourier methods. However, this requires some knowledge of the form of the electron distribution and this knowledge is not always readily available.

5. NEGATIVE ION LIFETIMES

As we have already seen, certain molecules form parent negative ions at low energies by a resonance capture process. Such ions may, in the absence of radiative or collision stabilization, re-emit the captured electron and return to the neutral state. However, if the average time between capture and such autodetachment is large enough, i.e. about 10 μsec, the charged species may be observable mass spectrometrically.

Edelson *et al.*[49] have shown that the ion SF_6^- from SF_6, formed at low electron energies, exhibits such behaviour and that the single SF_6^- ion peak normally observed in the negative ion mass spectrum consists of two components, one being the ion itself and the other being the neutral SF_6 formed as a result of autodetachment. Using a time-of-flight mass spectrometer, they have separated the two species by applying a retarding potential to the ion beam just before it reaches the ion detector. This has the effect of slowing down the ions, causing them to appear at a higher mass number while the neutrals are unaffected and appear at the same mass number as before.

By measuring the intensities of the ion N and of the ion plus neutral No, and by assuming an exponential decay law, they have calculated an autodetachment life-time τ. Here, τ is given by the relation

$$N = No\ \exp^{-t}/\tau,$$

where t is the drift time of the beam before separation of the ions and neutrals. In their experiments, t was about 18·5 μsec. For SF_6^- production from SF_6, τ was found to be about 10 μsec.

This is a long time when compared with the vibration period of the molecule and it has been suggested that the ion is formed in an excited state in such a way that the energy is distributed over the many internal degrees of freedom of the ion. The lifetime τ is then a measure of the time taken for this energy to be concentrated in a particular mode leading to autodetachment.

Following Edelson *et al.*,[49] Compton *et al.*,[50] have re-examined the lifetime of the parent negative ion from SF_6 and have extended the measurements to the parent ions $C_6H_5NO_2^-$ and $(CH_3CO)_2^-$. A modified Bendix Model 14–206 mass spectrometer (Fig. 13) was used to separate the ions and neutrals by applying a "flat-topped" retarding potential by means of the time-of-flight adjust control at about 170 cm from the ion source. Single-pulse counting techniques ensured equal detector sensitivity to both the negative ions and the neutrals and at the same time allowed low flight tube pressures ($\sim 10^{-7}$ torr) to be used, thus minimizing detachment and charge transfer as a result of intermolecular collisions. Electron detachment due to collision with grids in the flight path and spontaneous or collision-induced dissociations of the parent ion were found to be negligible.

The results obtained for six different ion energies from 2000 to 4000 eV are given in Table 2.

TABLE 2. *Lifetimes of temporary negative ions for various ion energies*

Ion energy (eV)	Lifetime (μsec)		
	$C_6H_5NO_2^-$	SF_6^-	$(CH_3CO)_2^-$
2000	31	24	11·8
2500	38	28	11·45
2750	38	26	
3000	38	27	
3500	43	24	
4000	42	24	

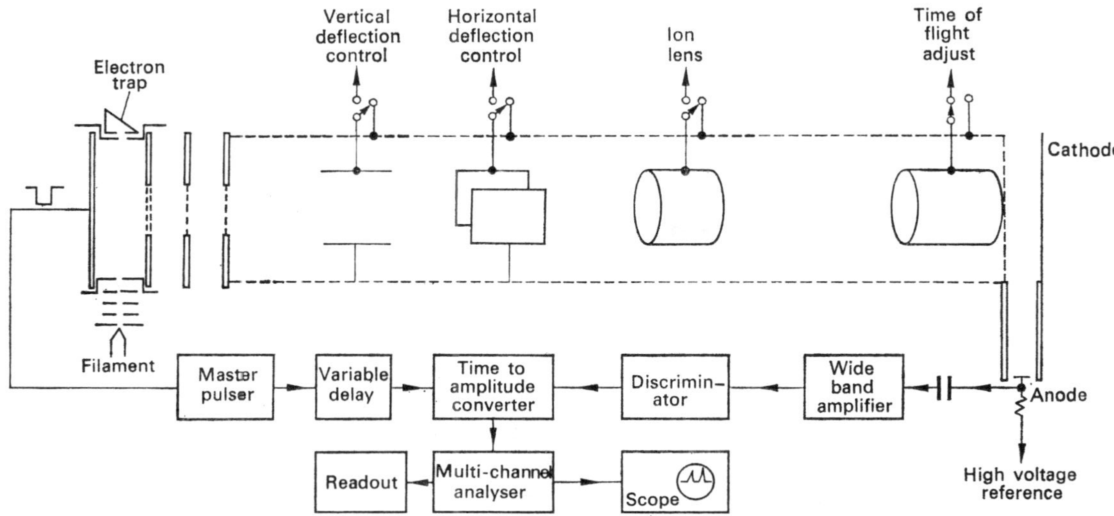

FIG. 13. Schematic diagram of the apparatus used in measuring negative ion lifetimes. (Diagram taken from ref. 50.)

Approximate mean values for the autodetachment lifetimes of SF_6^-, $C_2H_5NO_2^-$ and $(CH_3CO)_2^-$ are given as 25, 40 and 11·5 μsec respectively. The authors[(50)] speculate that the difference found in their measured lifetime of SF_6^-, and that of Edelson *et al.*[(49)] may be due to different focusing properties or source conditions in the two cases.

Clearly time-of-flight mass spectrometry offers a very powerful method for the determination of the lifetimes of short-lived species and it is likely that this will prove a fruitful line of investigation.

6. NEGATIVE ION–MOLECULE REACTIONS

The existence of negative ion–molecule reactions, i.e. the chemical reactions of negative ions in binary collisions with neutral species, has been known for some time but, in keeping with the whole field of negative ion studies, the amount of research devoted to the subject has been much more limited than the corresponding interest in positive ion–molecule reactions.[(51)]

In a review in 1963, Melton[(52)] reported that only about thirty such reactions had been examined and few have been added to the list since then. Reactions not included by Melton or studied since that review are presented in Table 3.

TABLE 3. *Some reactions of negative ions with neutral molecules*

System	Reaction	Reference
H_2O	$H_2O+H^- \rightarrow OH^-+H_2$	(53)
	$H_2O+O^- \rightarrow OH^-+OH$	(53)
NO_2	$O^-+NO_2 \rightarrow NO_2^-+O$	(54)
CS_2/SO_2 mixture	$CS^-+SO_2 \rightarrow SO_2^-+CS$	(55)
NH_3/SO_2 mixture	$NH_2^-+SO_2 \rightarrow SO_2^-+NH_2$	(55)
NH_3/CS_2 mixture	$NH_2^-+CS_2 \rightarrow CS_2^-+NH_2$	(55)
N_2O	$O^-+N_2O \rightarrow NO^-+NO$	(56)
	$O^-+N_2O \rightarrow O_2^-+N_2$	(56)
	$O^-+N_2O \rightarrow NO_2^-+N$	(56)

Experimentally the secondary product ions of such reactions may be identified by their pressure-dependence, since their intensity will depend upon the second power of the ion source pressure whereas the primary ion is linearly dependent upon the source pressure.

Positive ion–molecule reactions are known to play an important part in the radiation chemistry of many systems. The possible importance of negative ion–molecule reactions in such systems is not yet known although they might be expected to be significant, particularly for those systems containing atoms or molecules of high electron affinity.

7. NEGATIVE IONS PRODUCED BY SLOW SECONDARY ELECTRONS

Recently a new type of secondary negative ion has been found whose immediate precursor is not a primary negative ion but a slow secondary electron produced as the result of the collision of a fast electron with a neutral atom or molecule.[57] The intensities of such secondary negative ions show a squared dependence upon the total gas pressure in the ion source but the ionization efficiency curves of the secondary negative ion cannot be related to the curves for any primary ions in the system. Two molecules which have high electron capture cross-sections at low energies over a narrow energy band are SF_6 and CH_3I which form negative ions by the following processes:

$$SF_6 + e \longrightarrow SF_6^-$$

$$SF_6 + e \longrightarrow SF_5^- + F$$

$$CH_3I + e \longrightarrow I^- + CH_3$$

and these molecules have been used to demonstrate this effect.

Figures 14 and 15 show the ion currents for SF_6^- from SF_6 and I^- from CH_3I when these gases are ionized in a mass spectrometer in the presence of sulphur dioxide—the dotted line indicating the ion current observed[57] for the pure gases and the full line that in the gas mixture. It can be seen that certain peaks appear when the mixtures are ionized which do not appear in the pure gases. These peaks are due to secondary ion formation and the process can be explained as follows: an electron of initial kinetic energy E interacts with a sulphur dioxide molecule, excites this

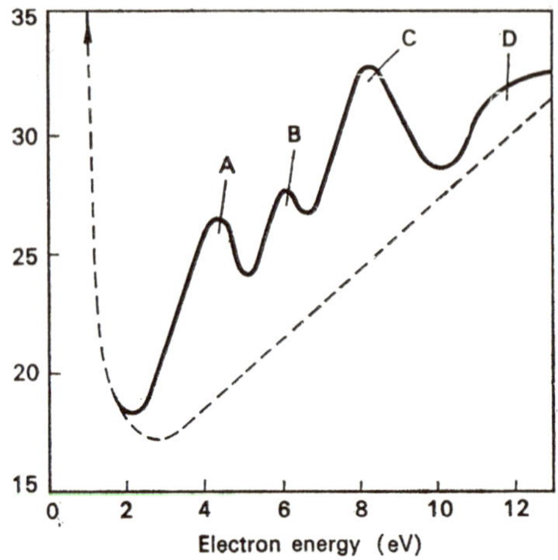

FIG. 14. Ion current due to the I^- ion from methyl iodide as a function of the electron accelerating voltage in pure methyl iodide (dotted line) and in a methyl iodide–sulphur dioxide mixture (full line). (Diagram taken from ref. 57.)

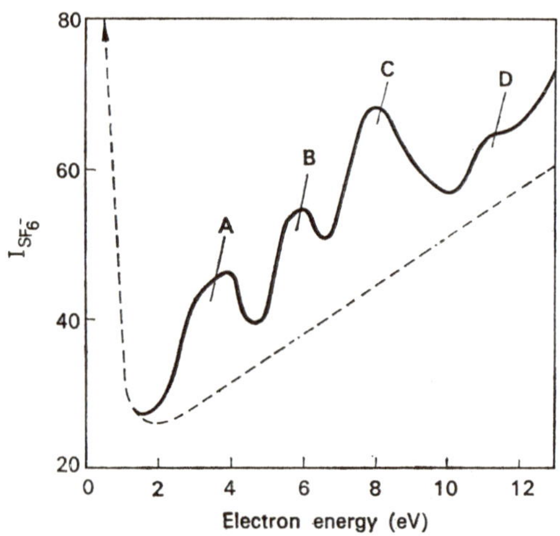

FIG. 15. Ion current due to the SF_6^- ion from sulphur hexafluoride as a function of the electron accelerating voltage in pure sulphur hexafluoride (dotted line) and in a sulphur hexafluoride–sulphur dioxide mixture (full line). (Diagram taken from ref. 57.)

molecule and emerges from the collision with a reduced kinetic energy E'. If this kinetic energy E' is in the correct range, the electron may be captured by the SF_6 or CH_3I molecule to form a negative ion. In SF_6/SO_2 gas mixtures the process is

$$SO_2 + e_E \longrightarrow SO_2^* + e_{E'}$$
$$e_{E'} + SF_6 \longrightarrow SF_6^-$$

and in CH_3I/SO_2

$$SO_2 + e_E \longrightarrow SO_2^* + e_{E'}$$
$$e_{E'} + CH_3I \longrightarrow I^- + CH_3,$$

where the asterisk indicates an excited molecule.

Gases which show this effect, e.g. SF_6 and CH_3I have been called "indicator" gases, and besides showing the occurrence of molecular excitation in the second component gas, they can also be used to demonstrate excitation of any positive ions produced and second or subsequent ionization levels.

It has also been suggested by Pignataro *et al.*[44] that since the capture peaks from certain metal carbonyls show a sharp energy dependence at energies higher than those for SF_6 and CH_3I, these carbonyls may be used as indicator gases for secondary electrons of higher kinetic energy.

8. APPLICATION OF THE QUASI-EQUILIBRIUM THEORY OF MASS SPECTRA TO NEGATIVE IONS

The need for a theoretical explanation of the general features observed in mass spectra, e.g. the occurrence of fragment ions formed from parent ions by bond dissociation and rearrangement, multiple charged ions, metastable ions, the shapes of ionization efficiency curves of both parent and fragment ions and temperature and isotope effects has resulted in the formulation of the quasi-equilibrium theory of mass spectra. This theory has been widely applied to positive ion studies (see, for example, ref. 58) and has proved to be of some value in the qualitative understanding of these phenomena but has only recently been applied to negative ion mass spectra.

In the original formulation[59, 60] of the quasi-equilibrium theory, two principal assumptions were made, namely that mass spectra are the result of series of competing, consecutive unimolecular decomposition reactions of excited parent ions and that the rates of each of these reactions can be calculated using absolute reaction rate theory. Absolute reaction rate theory is based on the assumption that reaction rate depends on the concentration and nature of an activated complex in equilibrium with the reactant species. From these assumptions it can be shown that the general expression for the average reaction rate of an ion is of the form

$$k(E, E') = \frac{1}{h}\left[\sum_{\varepsilon=0}^{E-E'} \varrho^*(\varepsilon)/\varrho(E)\right],$$

where E is the total internal energy of the system, E' the activation energy of the reaction and h is Planck's constant. The function $\varrho(E)$ represents the density of energy states of the reactant ion and $\varrho^*(\varepsilon)$ the density of states for the activated complex obtained by including all internal degrees of freedom except translation along the reaction coordinate. Because of the lack of experimental knowledge of these density of energy states functions, physical models have been used representing the internal degrees of freedom of the ion as a set of harmonic oscillators or as a set of harmonic oscillators plus a set of internal rotors.

The first application of this theory to negative ion mass spectra has been made by Winters and Kiser[61] using the negative ions produced by the electron bombardment of nickel tetracarbonyl, iron pentacarbonyl and the hexacarbonyls of chromium, molybdenum and tungsten. Using a Bendix Model 12–100 Time-of-Flight Mass Spectrometer, they have measured the relative abundances of the negative ions produced as a result of ion-pair processes at 70 eV and the energy of the dissociative capture maxima for each ion. The Model 12–100 has no electrical output, and spectra and capture maxima were examined by photographing the oscilloscope display. Relative abundances of the negative ions found at 70 eV are given in Table 4. No parent negative ions were found or any ions formed by rupture of the CO bond.

The energy of the dissociative capture maxima are given in Table 5.

It can be seen that as each CO group is removed there is a stepwise increase in the energy of the capture maximum. The occurrence of similar negative ions in the mass spectrum at 70 eV and in the dissociative capture energy range suggests that the decomposition scheme of the negative ions at 70 eV involves

TABLE 4. *Relative abundances of the negative ions in the mass spectrum of various metal carbonyls at 70 eV*

Ion	$Ni(CO)_4$	$Fe(CO)_5$	$Cr(CO)_6$	$Mo(CO)_6$	$W(CO)_6$
$M(CO)_5^-$			100	100	100
$M(CO)_4^-$		100	30	65	90
$M(CO)_3^-$	100	45	30	55	55
$M(CO)_2^-$	8	30	10	5	
$M(CO)^-$	3	5	5		

TABLE 5. *Energy of the dissociative capture maxima (eV)*

Ion	$Ni(CO)_4$	$Fe(CO)_5$	$Cr(CO)_6$	$Mo(CO)_6$	$W(CO)_6$
$M(CO)_5^-$			2·4	2·2	1·8
$M(CO)_4^-$		2·0	3·7	3·0	2·6
$M(CO)_3^-$	2·0	3·0	4·8	4·1	3·6
$M(CO)_2^-$	2·6	4·0	8·1	7·5	7·5
$M(CO)^-$	4·5	7·9	8·2	9·8	10·8
M^-	5·3	8·5	8·3	10·9	

successive removal of CO groups. Winters and Kiser[61] have proposed the following decomposition mechanism:

$$\begin{array}{lcl} M(CO)_x + e & \longrightarrow & M(CO)_{x-1}^- + CO^+ + e \\ & & \quad\downarrow \\ & & M(CO)_{x-2}^- + CO \\ & & \quad\downarrow \\ & & M(CO)_{x-3}^- + CO \\ & & \quad\downarrow \\ & & \quad\vdots \\ & & \quad\downarrow \\ & & M^- \qquad + CO \end{array}$$

i.e. the formation of a "pseudo-parent" $M(CO)_{x-1}^-$, since no parent negative ions were observed, and the subsequent elimination of CO groups by consecutive unimolecular reactions.

On the basis of this reaction mechanism, they have applied the quasi-equilibrium theory to predict the negative ion mass spectra expected as a result of ionization by electrons of 70 eV energy. The rate expression used was that of Vestal *et al.*[62] and activation energies for each reaction were taken as the energy difference between the dissociative capture maxima.

TABLE 6. *Calculated relative abundances of the negative ions in the mass spectrum of various metal carbonyls at* 70 *eV*

Ion	$Ni(CO)_4$	$Fe(CO)_5$	$Cr(CO)_6$	$Mo(CO)_6$	$W(CO)_6$
$M(CO)_5^-$			100	100	100
$M(CO)_4^-$		100	37	68	88
$M(CO)_3^-$	73	40	36	59	69
$M(CO)_2^-$	100	32	2	8	
$M(CO)^-$	7	1	2		
M^-	1				

The results obtained by using the theory are shown in Table 6. Comparison of Table 4 and Table 6 shows that the theoretically predicted relative abundances of the various negative ions agree very well with those found experimentally with the exception of the ions from nickel tetracarbonyl. This indicates that the decomposition mechanism used in the calculations is a reasonable one. To date, no further applications of the quasi-equilibrium theory to negative ion mass spectra has been made so that it is not yet possible to judge how successful it will be in dealing with such phenomena in other cases.

9. NEGATIVE ION MASS SPECTRA

The relative proportions of the different ions produced when a molecule is bombarded with ionizing electrons is usually referred to as the mass spectrum of the molecule. When the energy of the electrons is greater than about 30–40 eV, the relative abundances of the ions produced are generally independent of the energy and the positive ion mass spectra of many compounds have been obtained under such conditions. Negative ions cannot be similarly tabulated because of the different modes of formation of these ions. (At energies of $\sim <2$ eV they are formed primarily by electron capture, at $\sim$2–10 eV by dissociative resonance capture and at $\sim$70 eV by ion-pair formation.) Consequently the ions observed and their relative abundances are markedly dependent upon the electron energy. In addition, at high source pressures all three modes of formation may take place concurrently due to secondary processes and ion abundances may exhibit different pressure dependences. However, at 70 eV and source pressures of about 10^{-5} mmHg the predominant mode of formation is by ion-pair formation and in the following discussion (except where stated) we shall consider negative ions formed in a variety of halogenated compounds under these source conditions using a Bendix TOFMS, model 3015. Previous models of the Bendix TOFMS, e.g. 12–107 and 14–107, also enable rapid switching from the positive to the negative ionization mode, but on such instruments the analogue output system handles positive ions only, negative ions having to be studied on the oscilloscope output. On the 3015 model recording of negative ions by means of the analogue output system is made possible by floating the electron multiplier detector and the analogue electrometers, the ion source being operated at ground potential. Spectra were recorded using a 1 mV Kent potentiometric recorder. The same source pressure has been used for all compounds in order to reduce the differences in ionic species and their relative abundances due to possible ion–molecule reactions.

We have made no attempt to measure the appearance potentials of the ions observed or to evaluate the different possible processes leading to the formation of particular ions. The aim of this investigation was to study a variety of organic compounds in order to observe the general features or characteristics which determine the patterns of negative ion formation.

We have suggested, for each compound studied, on the basis of their positive and negative ion mass spectra, possible reactions involving ion-pair formation to account for the particular ions observed. These reactions are not exhaustive and in many cases are inadequate to explain ion formation. In addition the reaction energetics cannot be evaluated. Also since the end product is often a free electron, the energy of which may vary considerably, some of the slower secondary electrons may well experience capture leading to the formation of other ions. We have worked at relatively low source pressures in order to minimize the contributions of such secondary reactions.

In the tables we have expressed intensities of the negative ions observed relative to the abundance of the most extensively formed ion (usually Cl^- or F^-), and in the figures we have plotted the logarithm of the relative ion currents versus the mass number of the ion. For ions with a relative ion intensity of unity we have indicated a small positive value (instead of zero) on the diagrams so that all contributing ions may be seen.

Hexafluoroethane, C_2F_6

In Fig. 16 and tabulated below we show the relative abundances of the negative ions found in the mass spectrum of hexafluoroethane. Formation of the F^- and CF_3^- ions have been reported previously.(15) It is rather surprising that the $C_2F_5^-$ ion should be entirely absent from the mass spectrum since, in view of its stable electronic structure, it might have been expected to have been formed extensively. In the positive ion mass spectrum $C_2F_5^+$ is a fairly abundant ion, being formed to about 41% of the extent

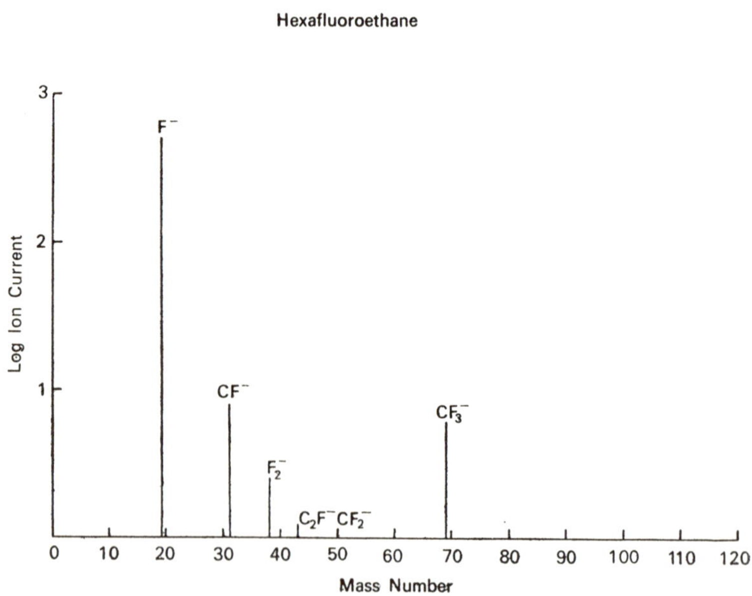

FIG. 16. Relative intensities of the negative ions formed by hexafluoroethane.

of the major ion CF_3^+. There is no evidence in the negative ion spectrum for ions of the type $C_2F_3^-$ or for any fragment ion containing two carbon atoms except for a small C_2F^- ion. This is in sharp distinction with data reported[4, 63] for ethane where the C_2^- and C_2H^- ions are the principal carbon-containing ions.

C_2F_6	F^-	CF^-	F_2^-	C_2F^-	CF_2^-	CF_3^-
Relative ion currents	1000	16	5	2	1	12

It is to be expected on the basis of their electronic structure that the F^- and CF_3^- ions will be formed extensively. In ethane although the H^- ion is abundant the CH_3^- ion is not observed, however the CF_3 radical almost certainly will have a much larger electron affinity than does the methyl radical.

Possible ion-pair formation processes which correlate the ions observed in both the positive and negative mass spectra are:

$$
\begin{aligned}
C_2F_6 + e &\longrightarrow F^- + C_2F_5^+ + e \\
&\longrightarrow CF^- + (CF_2 + 3\,F)^+ + e \\
&\longrightarrow CF^- + (CF_3 + 2\,F)^+ + e \\
&\longrightarrow F_2^- + CF_2^+ + CF_2 + e \\
&\longrightarrow F_2^- + C_2F_4^+ + e \\
&\longrightarrow C_2F^- + F^+ + 4\,F + e \\
&\longrightarrow CF_2^- + (CF_3 + F)^+ + e \\
&\longrightarrow CF_2^- + (CF_2 + 2\,F)^+ + e \\
&\longrightarrow CF_3^- + CF_3^+ + e
\end{aligned}
$$

We have bracketed certain species in some of the equations to indicate that either of the species may carry the positive charge.

The small extent of formation of the CF_2^- ion is initially rather unexpected but less so when it is remembered that in the positive ion spectrum of CF_4 the species CF_4^+ is virtually absent. In CF_4 the ionic species and order of formation are $CF_3^+ > CF_2^+ > CF^+$. Comparable species in hexafluoroethane are CF^- and C^-; the former is found quite abundantly but we have no evidence for the formation of C^-. It appears that negative ionic species leading to an odd number of outer electrons are formed to only very minor extents.

Fluoroform, CF_3H

Our data for fluoroform are shown in Fig. 17 and in the table below. Only the formation of the F^- ion has been reported previously. The most extensively formed ions are F^- and CF_3^- and the mass spectrum shows many similar features to that of hexafluoroethane, notably the appreciable formation of the CF^- and F_2^- ions with only a low-intensity peak due to CF_2^-. In

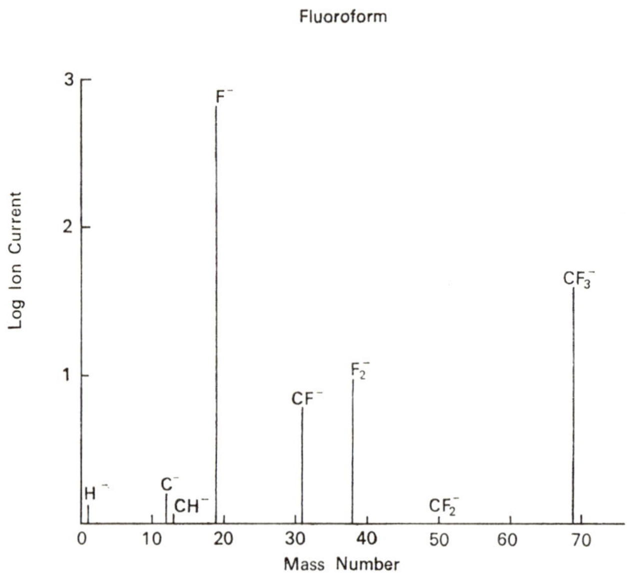

FIG. 17. Relative intensities of the negative ions formed by fluoroform.

addition H^- and CH^- are formed although neither of the ions are produced as abundantly as the analogous fluorine fragments. The C^- ion is also observed unlike in hexafluorethane.

CF_3H	H^-	C^-	CH^-	F^-	CF^-	F_2^-	CF_2^-	CF_3^-
Relative ion currents	2	2	1	1000	6	10	1	85

In the positive ion mass spectrum of fluoroform the most abundant ions are CF_3^+ and CF_2H^+, the latter being about 90% of the former. In the negative ion mass spectrum no fragments containing hydrogen and fluorine are observed although an ion such as CF_2H^- would have a stable electronic structure and, by analogy with CF_3^-, might have been expected to have been formed extensively.

Possible reactions leading the observed ions are:

$$
\begin{aligned}
CF_3H + e &\longrightarrow H^- + CF_3^+ + e\\
&\longrightarrow C^- + (HF + 2\,F)^+ + e\\
&\longrightarrow CH^- + F^+ + 2\,F + e\\
&\longrightarrow CF^- + (HF + F)^+ + e\\
&\longrightarrow F_2^- + CHF^+ + e\\
&\longrightarrow CF_2^- + HF^+ + e\\
&\longrightarrow CF_3^- + H^+ + e
\end{aligned}
$$

Carbon tetrachloride, CCl_4

Negative ion formation in carbon tetrachloride (Fig. 18) has been studied by several workers.[10–15] The ions we have observed are C^-, Cl^-, CCl^-, Cl_2^-, CCl_2^- and CCl_3^- and these are tabulated below. Fox and Curran report[11] that the Cl_2^- and CCl_3^- ions are about 450 times less abundant than the Cl^- ion; this is in reasonable agreement with our measurements that $Cl^- : Cl_2^- : CCl_3^- = 500 : 10 : 1$. We have observed CCl^- to be an abundantly formed ion, there appears to be no previous reference to the formation of this ion in carbon tetrachloride, and the ion appears to be formed much more extensively than is the CF^- ion from fluorocarbons.

Of particular interest is the extensive formation of both the C^- and CCl_2^- ions, since the results reported above for hexafluoroethane and fluoroform would indicate these, by analogy with CF_2^-, to be of minor importance. It appears that, for chlorinated hydrocarbons, the fragment ions are formed more abundantly and also that ions with odd-electron structures are formed more readily than in the case of fluorinated compounds. This may reflect the greater electron affinity of chlorine compared with fluorine.

CCl_4	C^-	Cl^-	CCl^-	Cl_2^-	CCl_2^-	CCl_3^-
Relative ion currents	2	1000	22	20	6	2

In the positive ion spectra of carbon tetrachloride the principal ions reported are Cl^+, CCl^+, CCl_2^+ and CCl_3^+. No molecular ion is observed.

$$
\begin{aligned}
CCl_4 + e &\longrightarrow Cl^- + CCl_3^+ + e \\
&\longrightarrow CCl^- + (3\ Cl)^+ + e \\
&\longrightarrow Cl_2^- + CCl_2^+ + e \\
&\longrightarrow CCl_2^- + (2\ Cl)^+ + e \\
&\longrightarrow CCl_3^- + Cl^+ + e
\end{aligned}
$$

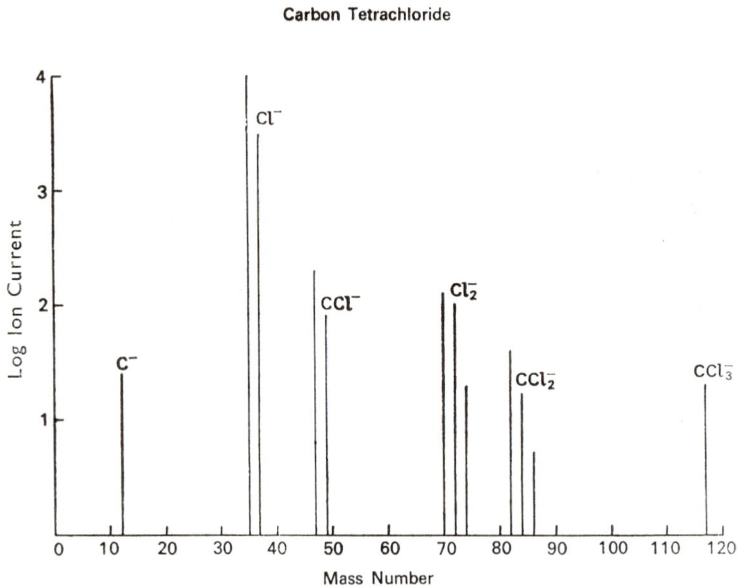

FIG. 18. Relative intensities of the negative ions formed by carbon tetrachloride.

Chloroform, $CHCl_3$

The principal ions observed in the negative ion mass spectrum of chloroform (Fig. 18) at 30 eV are C^-, CH^-, Cl^-, CCl^-, Cl_2^- and CCl_2^-. These are tabulated below. Ions previously reported were Cl^- and HCl_2^-. We found no evidence for HCl_2^- ion formation and have concluded that the ion was wrongly identified being confused with Cl_2^-. We confirmed that the ion was Cl_2^- by adding fluoroform to the system, using the ion CF_3^- as a mass-marker at $(m/e) = 69$. In Fig. 20 we show the negative ion mass spectrum in the region 69 to 75 a.m.u. When fluoroform is added it is clear that HCl_2^- is not a significant contributing ion, whereas Cl_2^- is.

$CHCl_3$	C^-	CH^-	Cl^-	CCl^-	Cl_2^-	CCl_2^-
Relative ion currents	1	6	1000	4	6	1

It is somewhat surprising that the ion CCl_3^- is absent from the spectrum although it was formed appreciably in carbon tetrachloride. Both C^- and CH^- are formed more abundantly than they were in fluoroform but, as with fluoroform, no ion is observed which contains carbon, hydrogen and the halogen. In the positive ion mass spectrum of chloroform Cl^+, CCl^+, and CCl_2^+ are the predominant ions, the CCl_3^+ ion is formed to only a minor extent, whereas in carbon tetrachloride it is the most abundant ion. In this sense the positive and negative ion mass spectra of chloroform exhibit a most interesting similarity in that although the CCl_3^+ and CCl_3^- ions are possible and are found abundantly in the fully halogenated compound, the ion with one less chlorine atom is the one formed in the case of chloroform.

The ions formed may be attributed to the following reactions:

$$
\begin{aligned}
CHCl_3 + e \longrightarrow\ & C^- + (HCl + 2\,Cl)^+ + e \\
\longrightarrow\ & CH^- + (3\,Cl)^+ + e \\
\longrightarrow\ & Cl^- + CHCl_2^+ + e \\
\longrightarrow\ & CCl^- + (HCl + Cl)^+ + e \\
\longrightarrow\ & Cl_2^- + CHCl^+ + e \\
\longrightarrow\ & CCl_2^- + HCl^+ + e
\end{aligned}
$$

Methylene dichloride, CH_2Cl_2

The negative ions found at 30 eV in the mass spectrum of methylene dichloride (Fig. 21) were C^-, CH^-, CH_2^-, Cl^-, CCl^- and Cl_2^- and are tabulated below. The relative intensities of the C^- and CH^- ions were about 1:6, the same ratio as observed in the case of chloroform. The identity of Cl_2^- (as distinct from HCl_2^-) was confirmed by the addition of fluoroform as a mass marker. No fragment ion was observed which

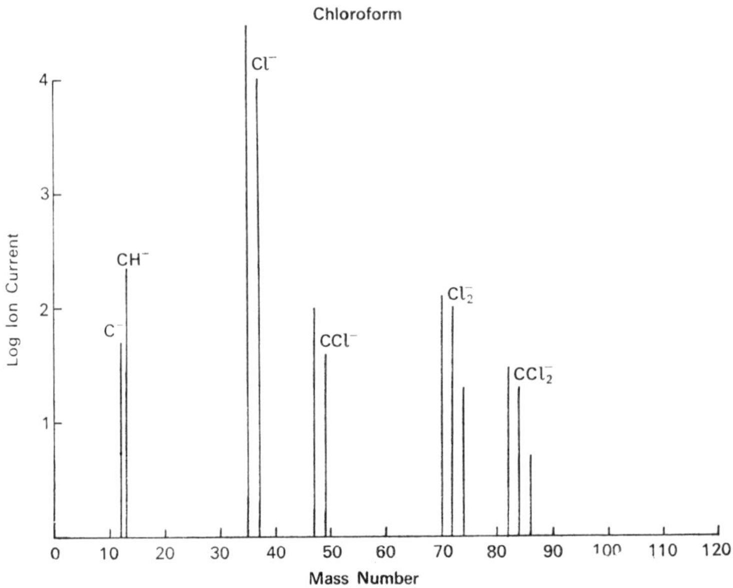

FIG. 19. Relative intensities of the negative ions formed by chloroform.

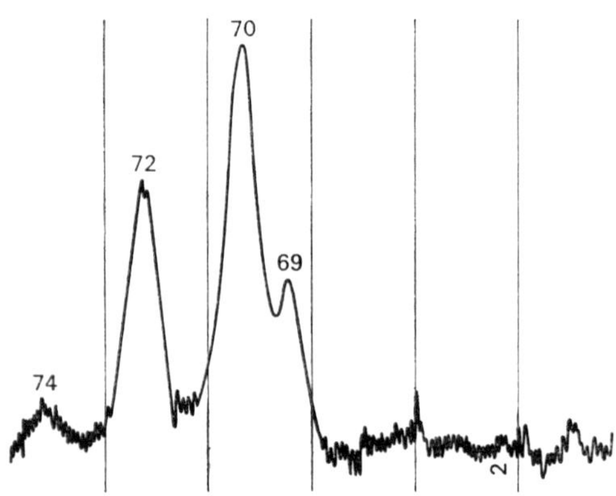

FIG. 20. Negative ion mass spectrum of chloroform in the 60–75 a.m.u. region. CF_3-peak due to added fluoroform.

contained carbon, hydrogen and the halogen which is in accord with the results reported above for fluoroform. The ion CCl_2^- was absent from the spectrum; this is somewhat surprising since it was found in the spectrum of chloroform. It appears that as the proportion of chlorine in the molecule decreases so the molecule shows less of a tendency to form large halogenated ions.

CH_2Cl_2	C^-	CH^-	CH_2^-	Cl^-	CCl^-	Cl_2^-
Relative ion currents	1	6	2	1000	2	21

In the positive ion mass spectrum $CHCl_2^+$ and CH_2Cl^+ are the major ions; these are analogous electronically to the quite abundant CCl^- ion found in this work.

$$
\begin{aligned}
CH_2Cl_2+e &\longrightarrow C^- +(2\,HCl)^+ +e \\
&\longrightarrow CH^- +(Cl+HCl)^+ +e \\
&\longrightarrow CH_2^- +(2\,Cl)^+ +e \\
&\longrightarrow Cl^- +CH_2Cl^+ +e \\
&\longrightarrow CCl^- +(Cl+HCl)^+ +e \\
&\longrightarrow Cl_2^- +CH_2^+ +e
\end{aligned}
$$

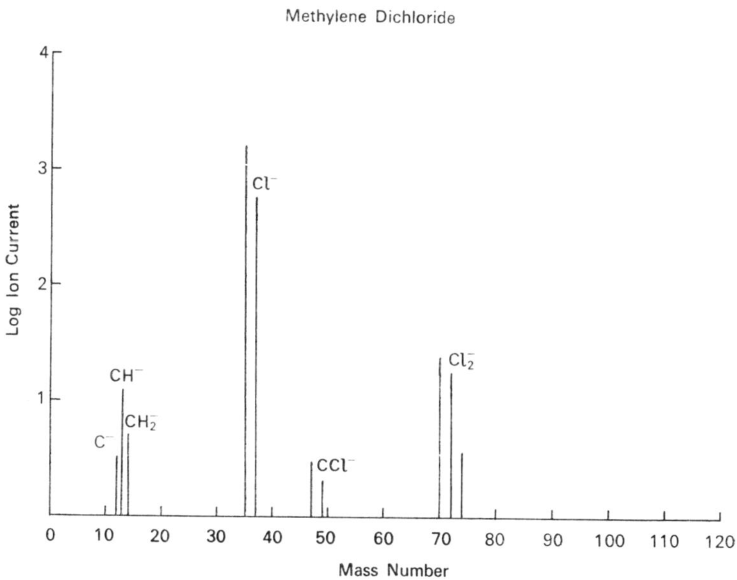

FIG. 21. Relative intensities of the negative ions formed by methylene chloride.

Methyl chloride, CH_3Cl

In Fig. 22 and in the table below we show the relative abundances of the ions found in the negative ion mass spectrum of methyl chloride using 30 eV ionizing electrons. The CH_3^- ion was not observed; however the ions, $CHCl^-$ and CH_2Cl^- were found in appreciable yields. This is in strong contrast to the results reported above for methylene chloride and chloroform where no ions containing carbon, hydrogen and chlorine together were observed. The H^- ion is also produced though our value for the relative intensity of the ion must be treated with some reserve in view of the possible mass discrimination towards such a light ion.

CH_3Cl	H^-	C^-	CH^-	CH_2^-	Cl^-	CCl_{35}^-	$CHCl_{35}^-$	CCl_{37}^- $+CH_2Cl_{35}^-$	$CHCl_{37}^-$
Relative ion currents	2	1	4	4	1000	1	2	3	1

CH_3Cl	$CH_2Cl_{37}^-$
Relative ion currents	1

In the positive ion mass spectrum CH_3^+, CH_2Cl^+ and CH_3Cl^+ are the principal ions together with a low intensity Cl^+ ion. The following reactions may be involved in the formation of negative ions:

$$\begin{aligned} CH_3Cl + e &\longrightarrow H^- + CH_2Cl^+ + e \\ &\longrightarrow C^- + (HCl + 2\,H)^+ + e \\ &\longrightarrow CH^- + (HCl + H)^+ + e \\ &\longrightarrow CH_2^- + HCl^+ + e \\ &\longrightarrow Cl^- + CH_3^+ + e \\ &\longrightarrow CCl^- + (3\,H)^+ + e \\ &\longrightarrow CHCl^- + (2\,H)^+ + e \\ &\longrightarrow CH_2Cl^- + H^+ + e \end{aligned}$$

Methyl iodide, CH_3I

The negative ions produced at 30 eV in the mass spectrum of methyl iodide (Fig. 23) are H^-, C^-, CH^-, CH_2^-, CH_3^- and I^-, and these are tabulated below. No ions such as CHI^- or

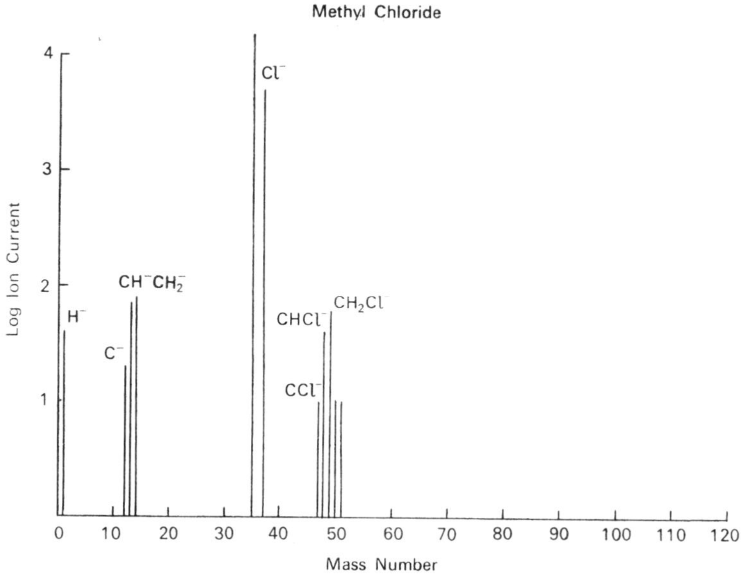

FIG. 22. Relative intensities of the negative ions formed by methyl chloride.

CH_2I^- were observed and, from the low intensity of the fragment ions, it appears that methyl iodide shows less tendency to form negative ions than does methyl chloride. This is not unexpected since the electron affinity of iodine is some 12 kcal mole^{-1} less than that of chlorine. An interesting difference also is the formation of the CH_3^- ion which is absent from the spectrum of methyl chloride.

CH_3I	H^-	C^-	CH^-	CH_2^-	CH_3^-	I^-
Relative ion currents	4	7	26	13	4	1000

In the positive ion mass spectrum ions such as CHI^+ and CH_2I^+ are produced quite abundantly.

$$
\begin{aligned}
CH_3I+e &\longrightarrow H^- + CH_2I^+ + e\\
&\longrightarrow C^- + (HI+2\,H)^+ + e\\
&\longrightarrow CH^- + (HI+H)^+ + e\\
&\longrightarrow CH_2^- + HI^+ + e\\
&\longrightarrow CH_3^- + I^+ + e\\
&\longrightarrow I^- + CH_3^+ + e
\end{aligned}
$$

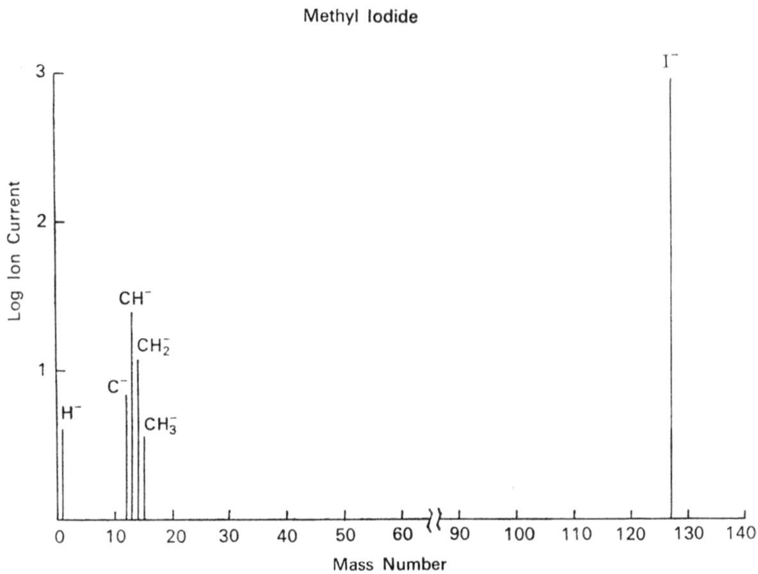

FIG. 23. Relative intensities of the negative ions formed by methyl iodide.

Symm.-tetrachloroethane, $CHCl_2$—$CHCl_2$

In Fig. 24 and tabulated below we report the data measured at 30 eV obtained for negative ion formation in symmetrical-tetrachloroethane. There is extensive negative ion production and the most obvious feature of the spectrum is that the molecule does not split symmetrically since the $CHCl_2^-$ ion is absent from the spectrum and the ion CCl_2^- is one of the less intense ions produced. The molecule when it fragments shows a marked tendency for the two carbon atoms to remain together, the ions C_2^-, C_2H^-, C_2Cl^- and $C_2Cl_3^-$ all being formed quite abundantly. An unusual ion found is the HCl_2^- which was not found in the corresponding spectra of chloroform or methylene dichloride. In certain respects fragmentation resembles that reported for ethane in that C^-, CH^-, C_2^- and C_2H^- are produced and the ratio C_2^-/CH^- is about 5.

The positive ion mass spectrum of symmetrical-tetrachloroethane indicates that the $CHCl_2^+$ ion, corresponding to a symmetrical split, is the major ion and also that there is a strong tendency for two-carbon fragment ions to be formed.

$CHCl_2$—$CHCl_2$	H^-	C^-	CH^-	C_2^-	C_2H^-	Cl^-
Relative ion currents	1	1	1	3	3	1000

CCl^-	C_2Cl^-	Cl_2	HCl_2^-	CCl_2^-	$C_2Cl_3^-$
2	3	49	6	1	23

Some of the possible reactions leading to negative ion formation under our experimental conditions may be:

$$
\begin{aligned}
CHCl_2\text{—}CHCl_2 + e &\longrightarrow H^- + C_2HCl_3 + e\\
&\longrightarrow C^- + (CHCl_2 + HCl + Cl)^+ + e\\
&\longrightarrow CH^- + (CHCl_2 + Cl_2)^+ + e\\
&\longrightarrow C_2^- + (HCl_2 + HCl_2)^+ + e\\
&\longrightarrow C_2H^- + (Cl_2 + HCl_2)^+ + e\\
&\longrightarrow Cl^- + (CHCl_2 + CHCl)^+ + e\\
&\longrightarrow CCl^- + (CHCl + CHCl_2)^+ + e\\
&\longrightarrow C_2Cl^- + (HCl + Cl_2)^+ + e\\
&\longrightarrow Cl_2^- + (C_2H_2Cl_2)^+ + e\\
&\longrightarrow HCl_2^- + C_2HCl_2^+ + e\\
&\longrightarrow CCl_2^- + CH_2Cl_2^+ + e\\
&\longrightarrow C_2Cl_3^- + (HCl + H)^+ + e
\end{aligned}
$$

Unsymm.-tetrachloroethane, CCl_3—CH_2Cl

The negative ion mass spectrum of unsymmetrical-tetrachloroethane (Fig. 25) is, somewhat surprisingly, much simpler than that of the symmetrical compound, only eight ionic species being observed compared with twelve in the case of $CHCl_2$—$CHCl_2$, and in general they are of lower intensity. No fragment ion containing two carbon atoms together with a halogen atom is observed although the C^-, CH^-, CH_2^-, C_2^- and C_2H^- ions are produced, the latter two ions being easily the most abundant. The CCl^- ion is formed but there is no evidence for the formation of ions such as CCl_3^-, CCl_2^- or HCl_2^-. The ion $C_2Cl_3^-$ is not found; this is considerably different from the symmetric compound where it was formed extensively. The absence of the

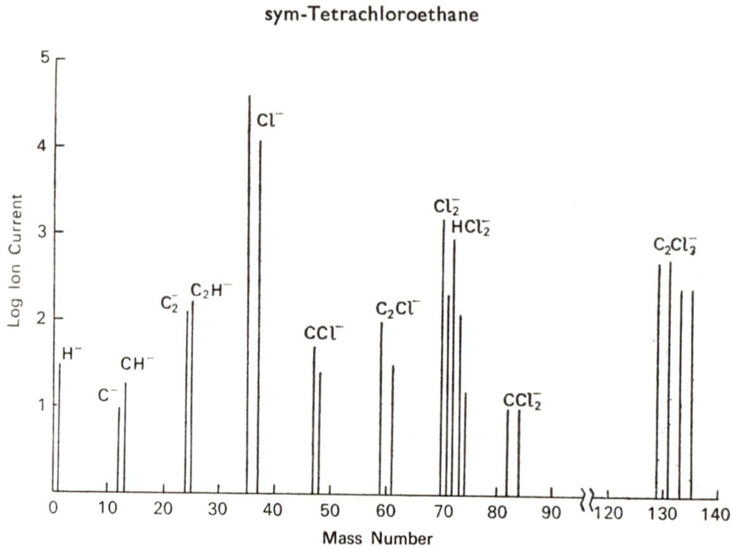

FIG. 24. Relative intensities of the negative ions formed by symmetrical-tetrachloroethane.

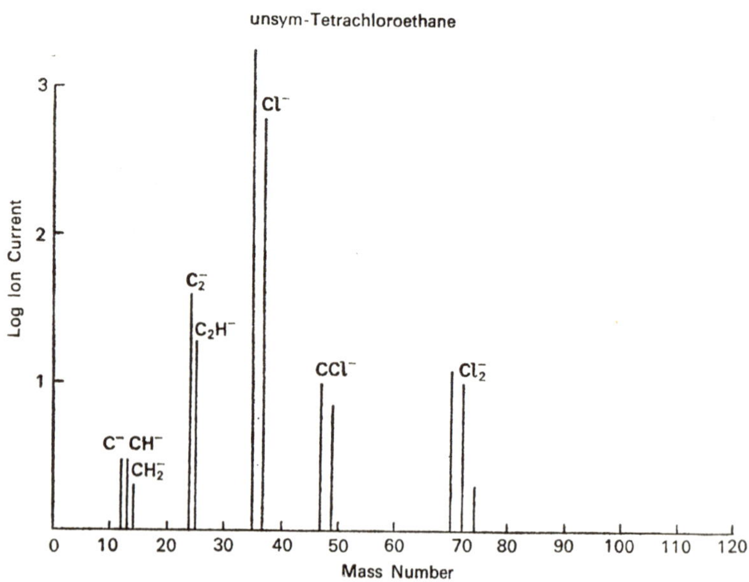

FIG. 25. Relative intensities of the negative ions formed by unsymmetrical-tetrachloroethane.

ions CCl_3^- and CH_2Cl^- together with the C_2-type fragments suggests that there is little symmetrical splitting. It is difficult to see why ions such as C_2Cl^- and CCl_2^- are not formed in the case of the unsymmetrical molecule; possibly their absence reflects the greater tendency of symmetrical molecules to form large negative fragment ions.

CCl_3–CH_2Cl	C^-	CH^-	CH_2^-	C_2^-	C_2H^-	Cl^-	CCl^-	C_2^-
Relative ion currents	1	1	1	13	8	1000	7	10

In the positive ion mass spectrum the major ions are fragment ions containing two carbon atoms, such as $C_2Cl_3^+$, $C_2HCl_2^+$, C_2Cl^+ and C_2H^+ together with a large CCl_3^+ ion which indicates some carbon–carbon bond fission.

$$
\begin{aligned}
CCl_3\text{–}CH_2Cl + e &\longrightarrow C^- + (CCl_3 + H_2Cl)^+ + e\\
&\longrightarrow CH^- + (CCl_3 + HCl)^+ + e\\
&\longrightarrow CH_2^- + (CCl_3 + Cl)^+ + e\\
&\longrightarrow C_2^- + (HCl + 2\ Cl)^+ + e\\
&\longrightarrow C_2H^- + (HCl + 3\ Cl)^+ + e\\
&\longrightarrow Cl^- + (CH_2 + CCl_3)^+ + e\\
&\longrightarrow Cl^- + (CCl_2 + CH_2Cl)^+ + e\\
&\longrightarrow CCl^- + (2\ Cl + CH_2Cl)^+ + e\\
&\longrightarrow CCl^- + (CCl_3 + 2\ H)^+ + e\\
&\longrightarrow Cl_2^- + (CCl + CH_2Cl)^+ + e\\
&\longrightarrow Cl_2^- + (CCl_2 + CH_2)^+ + e
\end{aligned}
$$

Tetrachloroethylene, $CCl_2{=}CCl_2$

In Fig. 26 we show the negative ion mass spectrum of tetrachloroethylene measured at 30 eV. It is apparent that the molecule is an abundant source of negative ions and, rather unusually, shows a large molecular ion, the only example of parent ion formation (with the exception of hexafluoroacetone which is discussed below) found in this study. This is presumably because the molecule-ion is stabilized with respect to decomposition by virtue of its symmetry and the strong carbon–carbon bonding. No CCl_2^- ion is observed in the spectrum suggesting that there is little symmetrical bond fission; this is supported by the prevalence of quite intense two carbon atom-containing fragment ions such as C_2^-, C_2Cl^- and $C_2Cl_4^-$. Presumably this is attributable to the presence of the double bond and suggests that interesting fragment ions may be observed for molecules containing conjugated olefinic linkages. The intensity of the C_2^- and C_2Cl^- ions is very much greater than observed in the case of symmetrical-tetrachloroethane.

$CCl_2{=}CCl_2$	C^-	C_2^-	Cl^-	CCl^-	C_2Cl^-	Cl_2^-	$C_2Cl_4^-$
Relative ion currents	1	25	1000	3	22	9	8

Comparison of negative ion formation by this molecule with that in ethane, ethylene and acetylene is interesting. In each of the latter three cases C_2^- and C_2H^- ions are formed, but the intensity of these fragments increases as the bonding between the carbon atoms getss tronger.

$$
\begin{aligned}
CCl_2{=}CCl_2 + e &\longrightarrow C^- + (2\,Cl + CCl_2)^+ + e\\
&\longrightarrow C_2^- + (4\,Cl)^+ + e\\
&\longrightarrow Cl^- + (CCl + CCl_2)^+ + e\\
&\longrightarrow CCl^- + (CCl_2 + Cl)^+ + e\\
&\longrightarrow C_2Cl^- + (3\,Cl)^+ + e\\
&\longrightarrow Cl_2^- + (C + CCl_2)^+ + e\\
&\longrightarrow Cl_2^- + (2\,CCl)^+ + e\\
&\longrightarrow C_2Cl_4^-
\end{aligned}
$$

Hexafluoroacetone, CF_3COCF_3

The negative ion mass spectrum of hexafluoroacetone (Fig. 27) consists of the following ions: O^-, F^-, CF^-, F_2^-, C_2F^-, CF_2^-, CF_3^- and $CF_3COCF_3^-$ and these are tabulated below. The parent ion is formed quite abundantly presumably reflecting the large electron affinities of oxygen and fluorine together with the multiple bonding in the carbonyl group. There is a two-carbon atom ion, C_2F^-, but no larger fragment ion probably because of their low electron affinity. In this respect it is similar to hexafluoroethane where a low-intensity C_2F^- ion was the largest two-carbon atom fragmentation ion. The actual intensity of the O^- ion

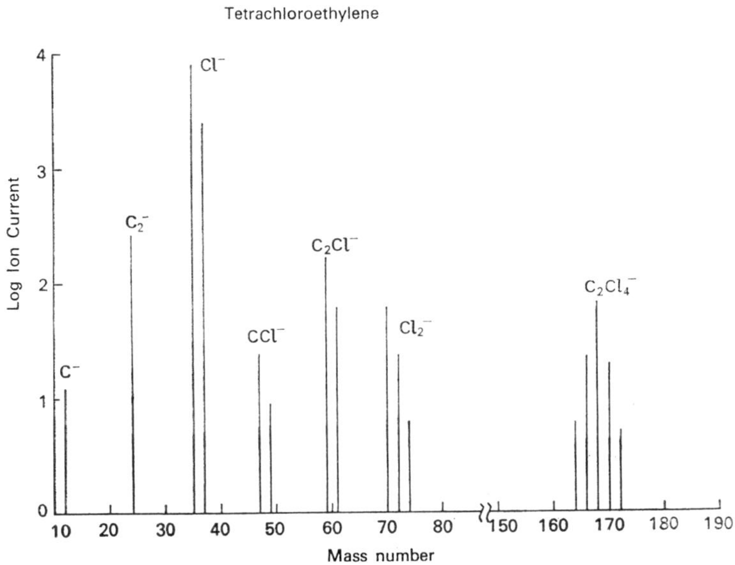

FIG. 26. Relative intensities of the negative ions formed by tetrachloroethylene.

is doubtful because of the background contribution to this ion.

CF_3COCF_3	O^-	F^-	CF^-	F_2^-	C_2F^-	CF_2^-	CF_3^-	$CF_3COCF_3^-$
Relative ion currents	530	1000	6	6	6	3	38	31

In the positive ion mass spectrum of hexafluoroacetone, CF_3^+ is the most abundant ion, other ions being formed including CF_3CO^+ and a small molecular ion. By analogy the ion CF_3CO^- might be found but we did not observe its formation.†

$$
\begin{aligned}
CF_3COCF_3 + e &\longrightarrow O^- + (C_3F_6)^+ + e \\
&\longrightarrow F^- + (CF_2 + CF_3CO)^+ + e \\
&\longrightarrow CF^- + (CF_3CO + 2\,F)^+ + e \\
&\longrightarrow F_2^- + (CF + CF_3CO)^+ + e \\
&\longrightarrow C_2F^- + (COF_5)^+ + e \\
&\longrightarrow CF_2^- + (CF_3CO + F)^+ + e \\
&\longrightarrow CF_3^+ + CF_3CO^+ + e \\
&\longrightarrow CF_3COCF_3^-
\end{aligned}
$$

† *Note added in proof:* in recent work at low electron energies we have observed this ion to be formed in low abundance.

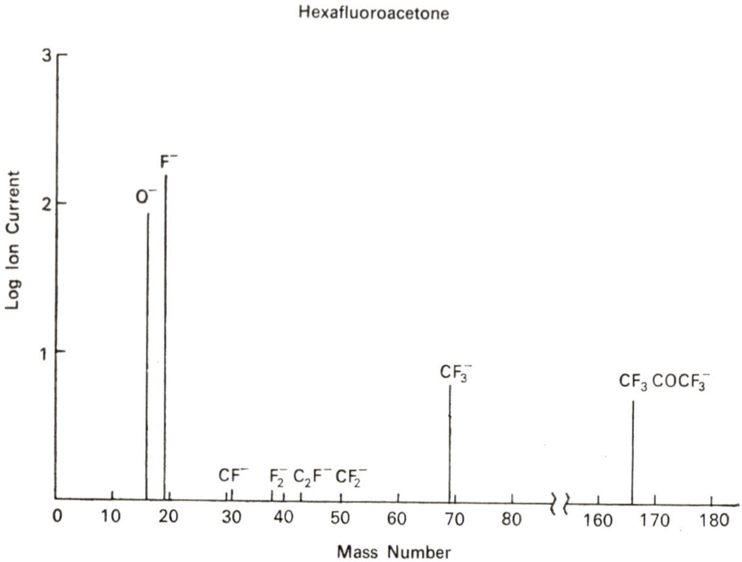

FIG. 27. Relative intensities of the negative ions formed by hexafluoroacetone.

Sulphur hexafluoride, SF_6

It has been shown previously that the formation of the SF_6^- ion at low electron energies can be used to indicate the spread of the energy of the ionizing electron beam and also as a marker or calibrant for the energy scale. In Fig. 28 we indicate the negative ions found in the mass spectrum of sulphur hexafluoride at 70 eV and these are also tabulated below. It is apparent that although SF_6^- and F^- are the most abundant species at this energy several other ions, including SF^- and SF_2^- are observed; these latter two ions have not been observed by other workers.

SF_6	F^-	S^-	F_2^-	SF^-	SF_2^-	SF_3^-	SF_4^-	SF_5^-	SF_6^-
Relative ion currents	160	1	1	6	1	1	1	14	1000

On the basis of their electronic structure it is to be expected that the SF^- and SF_5^- ions will be formed abundantly and this is experimentally verified; however it is rather surprising that the SF_3^- ion is of such low intensity.

The positive ion mass spectrum of sulphur hexafluoride has been measured by Dibeler and Mohler[65] who report that the most abundant ions at 70 eV are, in decreasing order of formation, SF_5^+, SF_3^+, SF^+, SF_4^+, S^+ and SF_2^+. No parent ion was observed. In addition several doubly-charged ions, e.g. SF_4^{++}, were found.

Possible reactions leading to ion-pair formation are:

$$
\begin{aligned}
SF_6+e &\longrightarrow SF_5^- + F^+ + e\\
&\longrightarrow SF_4^- + F^+ + F + e\\
&\longrightarrow SF_3^- + F^+ + 2\,F + e\\
&\longrightarrow SF_2^- + F^+ + 3\,F + e\\
&\longrightarrow SF^- + F^+ + 4\,F + e\\
&\longrightarrow S^- + F^+ + 5\,F + e\\
&\longrightarrow F_2^- + SF_4^+ + e\\
&\longrightarrow F^- + SF_5^+ + e
\end{aligned}
$$

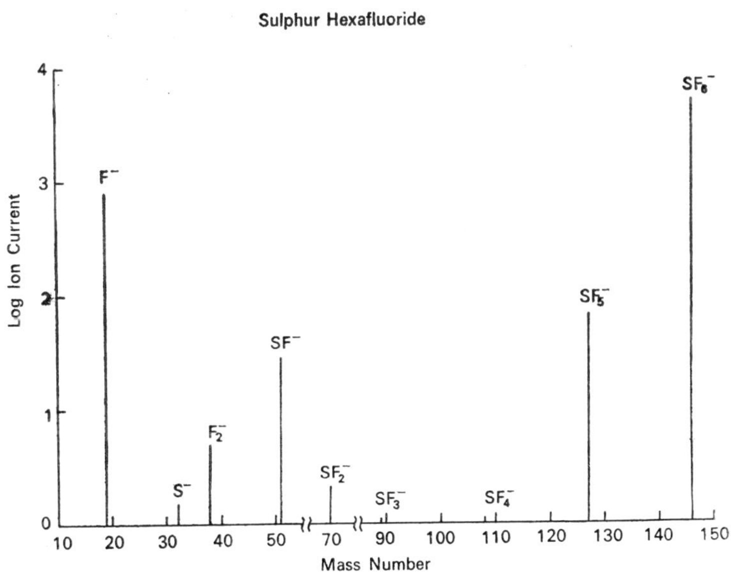

FIG. 28. Relative intensities of the negative ions formed by sulphur hexafluoride.

The results reported above enable correlations to be made between the molecules studied and also some general patterns of negative ion formation to be deduced. In all cases where a halogen atom is present in the molecule this is the most abundantly formed negative ion. Somewhat surprisingly formation of the halogen molecule negative ion is frequently quite extensive particularly in the case of chlorine and this probably reflects the greater electron affinity of chlorine.

(a) CF_3H *and* CCl_3H

Below we have compared the negative ions formed in chloroform and fluoroform giving the most abundant ion in each spectrum the same intensity.

Ion	C^-	CH^-	X^-	CX^-	X_2^-	CX_2^-	CX_3^-
CF_3H	2	1	1000	6	10	1	85
CCl_3H	1	6	1000	4	6	1	—

The general pattern of negative ion formation in these two compounds is similar with the notable exception of the CX_3^- ion which is not formed in the case of chloroform. The CX^- ion is formed more abundantly when X = F and the CX_2^- is formed to only a small extent. No ion such as CHX^- is formed for either compound suggesting that such a species has a low electron affinity.

(b) CH_3Cl *and* CH_3I

The main difference between these molecules is that the iodide forms no ions of the type CX^-, CHX^- and CH_2X^-, however more of the carbon–hydrogen ions appear to be formed in the case of the iodide, presumably because of the easy elimination of hydrogen iodide by reactions such as

$$CH_3I + e \longrightarrow CH^- + (H + HI)^+ + e$$

Another interesting difference is the observation that the CH_3^- ion is formed in the case of the iodide only.

Ion	H^-	C^-	CH^-	CH_2^-	CH_3^-	X^-	XC^-	CHX^-	CH_2X^-
CH_3Cl	2	1	4	4	—	1000	1	2	3
CH_3I	4	7	26	13	4	1000	—	—	—

(c) CCl_4, CCl_3H, CCl_2H_2 *and* $CClH_3$

Ion	H^-	C^-	CH^-	CH_2^-	Cl^-	CCl^-	$CHCl^-$	CH_2Cl^-	Cl_2^-	CCl_2^-	CCl_3^-
CCl_4	—	2	—	—	1000	22	—	—	20	6	2
CCl_3H	—	1	6		1000	4	—	—	6	1	—
CCl_2H_2	—	1	6	2	1000	2	—	—	21	—	—
$CClH_3$	2	1	4	4	1000	1	2	3	—	—	—

A most interesting trend exhibited by this series of molecules is that the largest observed chlorine-containing ion contains one chlorine atom less than the maximum possible number. For example, although the CCl_3^- ion is a possible product ion from chloroform the largest observed ion is CCl_2^-. Similarly, in methylene dichloride, where the CCl_2^- might be expected, the largest ion formed is CCl^- and in methyl chloride the CCl^- ion is of very low intensity. It appears that decrease in the halogen content of the molecule is accompanied by a decreasing tendency to form large fragment ions.

In general ions of the type $CHCl^-$ are not formed and ions such as CH^- are, not surprisingly, produced to a lesser extent than their chlorinated counterparts.

(d) $CHCl_2$—$CHCl_2$, CH_2Cl—CCl_3 *and* CCl_2=CCl_2

The most notable feature common to this group of molecules is a tendency not to fragment symmetrically by carbon–carbon bond fission but for the two carbon atoms to be present in the fragment ions. The absence of the CCl_2^- ion in two cases and its minor formation in the third suggests that it is not a very stable negative ion. This is confirmed by its low intensity of formation from compounds such as those discussed in (c). The formation of the molecular ion in this series is peculiar to tetrachloroethylene.

Ion	C^-	CH^-	CH_2^-	C_2^-	C_2H^-	Cl^-
$CHCl_2$—$CHCl_2$	1	1	—	3	3	1000
CCl_3—CH_2Cl	1	1	1	13	8	1000
CCl_2=CCl_2	1	—	—	25	—	1000

Ion	CCl^-	C_2Cl^-	Cl_2^-	HCl_2^-	CCl_2^-	$C_2Cl_3^-$	$C_2Cl_4^-$
$CHCl_2$—$CHCl_2$	2	3	49	6	1	23	—
CCl_3—CH_2Cl	7	—	10	—	—	—	—
CCl_2=CCl_2	3	22	9	—	—	—	8

From this rather limited investigation it appears that, for the halogenated molecules studied, negative ion formation is more extensive in compounds containing chlorine atoms than in those containing fluorine or iodine and that fragmentation tends to favour the formation of ions containing an odd number of halogen atoms and not to produce "mixed" ions such as CHX^-.

The generation of negative ions is clearly influenced by the size, symmetry and bonding of the molecule being bombarded by electrons. In general, large, symmetrical multiple-bonded molecules appear to undergo less fragmentation and consequently form larger negative ions, sometimes including the parent ion, than molecules without these characteristics.

The value of the time-of-flight mass spectrometer in studying the attachment of electrons to molecules over a wide range of electron energy and its potentialities in the evaluation of the appearance potentials and the lifetimes of the negative ions formed is clear. The application of the technique to the rather neglected field of negative ion-molecule reactions, taking advantage of the field-free operating conditions of the ion source, is another obvious area in which useful results may accrue. In addition information regarding the kinetic energy of negative ions and also for inelastic electron–molecule interactions other than attachment should be readily obtainable.

REFERENCES

1. Thomson, J. J., *Proc. Roy. Soc.* **89,** 1 (1913).
2. Bibliography in *Advances in Mass Spectrometry*, edited by W. L. Mead (Institute of Petroleum 1966), p. 1019.
3. Chanin, L. M., Phelps, A. V., and Biondi, M. A., *Phys. Rev. Letters* **2,** 344 (1959).
4. Melton, C. E., and Rudolph, P. S., *J. Chem. Phys.* **31,** 1485 (1959).
5. Calcote, H. F., Kurzius, S. C., and Miller, W. J., *10th Symposium (International) on Combustion, 1965*, **10,** 605.

6. Gaines, A. F., Kay, J., and Page, F. M., *Trans. Faraday Soc.* **62,** 874 (1966).
7. von Ardenne, M., and Tümmler, R., *Naturwiss* **45,** 414 (1958).
8. Rapp, G. A., and Melton, C. E., *J. Am. Chem. Soc.* **80,** 3509 (1958).
9. Massey, H. W., *Negative Ions* (Cambridge University Press, 1950).
10. Buchel'nikova, I. S., *J. Expt. Theoret. Phys. (USSR)*, **35,** 1119 (1958).
11. Fox, R. E., and Curran, R. K., *J. Chem. Phys.* **34,** 1595 (1961).
12. Hickam, W. M., and Berg, D., in *Advances in Mass Spectrometry*, edited by J. D. Waldron (Pergamon Press, 1959), p. 458.
13. Baker, R., and Tate, J., *Phys. Rev.* **53,** 683 (1938).
14. Craggs, J. D., McDowell, C. A., and Warren, J., *Trans. Faraday Soc.* **48,** 1093 (1952).
15. Reese, R., Dibeler, V. H., and Mohler, F. L., *J. Res. Nat. Bur. Standards* **57,** 367 (1956).
16. Marriott, J., Thorburn, R., and Craggs, J. D., *Proc. Phys. Soc. (London)* **B67,** 437 (1954).
17. Clarke, E. M., *Can. J. Phys.* **32,** 764 (1954).
18. Marmet, P., and Kerwin, L., *J. Appl. Phys.* **31,** 2071 (1960).
19. Marmet, P., and Kerwin, L., *Can. J. Phys.* **38,** 787 (1960).
20. Fox, R. E., Hickam, W. M., Grove, D. J., and Kjeldaas, T., Jr., *Rev. Sci. Instr.* **26,** 1011 (1955).
21. Melton, C. E. and Hamill, W. H., Notre Dame Report, University of Notre Dame, Indiana, 1963.
22. Melton, C. E., and Hamill, W. H., *J. Chem. Phys.* **41,** 546 (1964).
23. Schultz, G. J., *Phys. Rev.* **113,** 816 (1959).
24. Frost, D. C., and McDowell, C. A., *Can. J. Chem.* **38,** 407 (1960).
25. Frost, D. C., and McDowell, C. A., *J. Chem. Phys.* **29**, 964 (1958).
26. Fox, R. E., *Phys. Rev.* **100,** 1229 (1957).
27. Fox, R. E., *Phys. Rev.* **109,** 2008 (1958).
28. Frost, D. C., and McDowell, C. A., *J. Chem. Phys.* **29,** 503 (1958).
29. Fox, R. E., *J. Chem. Phys.* **26,** 1281 (1957).
30. Frost, D. C., and McDowell, C. A., *J. Chem. Phys.* **29**, 1424 (1958).
31. Curran, R. K., *J. Chem. Phys.* **35,** 1849 (1961).
32. Kraus, K. A., *Z. Naturf.* **16a,** 1378 (1961).
33. Fox, R. E., *J. Chem. Phys.* **32,** 285 (1960).
34. Curran, R. K., and Fox, R. E., *J. Chem. Phys.* **34,** 1590 (1961).
35. Halmann, M., and Klein, Y., *J. Chem. Soc.* 4324 (1964).
36. Halmann, M., and Klein, Y., in *Advances in Mass Spectrometry*, edited by W. L. Mead (Institute of Petroleum, 1964), p. 267.
37. Grob, V. E., *J. Chem. Phys.* **39**, 972 (1963).
38. Fox, R. E., and Curran, R. K., *J. Chem. Phys.* **34,** 1595 (1961).
39. Curran, R. K., *J. Chem. Phys.* **34,** 2007 (1961).
40. Hickam, W. M., and Fox, R. E., *Phys. Rev.* **98,** 557a (1955).
41. Hickam, W. M., and Fox, R. E., *J. Chem. Phys.* **25**, 642 (1956).
42. Curran, R. K., *J. Chem. Phys.* **34,** 1069 (1961).
43. Asundi, R. K., and Craggs, J. D., *Proc. Phys. Soc. (London)* **83,** 611 (1964).
44. Pignataro, S., Foffani, A., Grasso, F., and Cantone, B., *Z. Physik. Chem.* **47**, 106 (1965).
45. Marmet, P., *Can. J. Phys.* **42,** 2102 (1964).

46. HARTWIG, D. H., and ULMER, K., *Z. Physik* 1963, **73**, 294.
47. WINTERS, R. E., COLLINS, J. H., and COURCHENE, W. L., *J. Chem. Phys.* **45**, 1931 (1966).
48. MORRISON, J. D., *J. Chem. Phys.* **39**, 200 (1963).
49. EDELSON, D., GRIFFITHS, J. E., and MCAFEE, K. B., JR., *J. Chem. Phys.* **37**, 917 (1962).
50. COMPTON, R. N., CHRISTOPHOROU, L. G., HURST, G. S., and REINHARDT, P. W., *J. Chem. Phys.* **45**, 4634 (1966).
51. Ion-molecule reactions in the gas phase, *Advances in Chemistry*, 1966, **58**.
52. MELTON, C. E. in *Mass Spectrometry of Organic Ions*, edited by F. W. McLafferty (Academic Press, New York and London, 1963), p. 163.
53. COTTIN, M. J., *Chemie Phys.* **56**, 1024 (1959).
54. CURRAN, R. K., *Phys. Rev.* **125**, 910 (1962).
55. KRAUS, K. A. MUELLER-DUYSING, W., and NEUERT, H., *Z. Naturf.* **16a**, 1385 (1961).
56. BURTT, B. P., and HENIS, J., *J. Chem. Phys.* **41**, 1510 (1964).
57. JACOBS, G., and HENGLEIN, A., in *Advances in Mass Spectrometry*, edited by W. L. Mead (Institute of Petroleum, 1966), p. 287.
58. ROSENSTOCK, H. M., and KRAUSS, M., in *Mass Spectrometry of Organic Ions*, edited by F. W. McLafferty (Academic Press, New York and London, 1963), p. 1.
59. ROSENSTOCK, H. M., Ph.D. thesis, University of Utah, Salt Lake City, 1952.
60. ROSENSTOCK, H. M., WALLENSTEIN, M. B., WAHRHAFTIG, A. L., and EYRING, H., *Proc. Nat. Acad. Sci. US* **38**, 667 (1952).
61. WINTERS, R. E., and KISER, R. W., *J. Chem. Phys.* **44**, 1964 (1966).
62. VESTAL, M. L., WAHRHAFTIG, A. L., and JOHNSTON, W. H., *J. Chem. Phys.* **37**, 1276 (1962).
63. BAILEY, T. L., MCGUIRE, J. M., and MUSCHLITZ, E. E., *J. Chem. Phys.* **26**, 568 (1957).
64. DORMAN, F. H., *J. Chem. Phys.* **44**, 3857 (1966).
65. DIBELER, V. H., and MOHLER, F. L., *J. Res. Nat. Bur. Standards*, **40**, 25 (1948).

THE REACTIONS OF IONS WITH MOLECULES IN THE GAS PHASE AND THE SPUTTERING OF SOLID SURFACES BY ION BOMBARDMENT

D. W. THOMAS

Chemistry Department, The University of Birmingham, Birmingham 15

INTRODUCTION

We have made a few modifications and additions[1-4] to a Bendix time-of-flight mass spectrometer principally in order that two quite different types of ion–molecule interactions may be studied, namely (a) gas phase collisions, and (b) the sputtering of solid surfaces. A subsidiary but related interest has involved the substitution of fast-moving neutrals (atoms or molecules) for the reactant ions in both the above types of interaction. Much of the auxiliary equipment required and alterations to the basic instrument are common to both types of study. The primary experimental requirement in each case is the generation of a stable beam of specified primary ions, with a chosen translational energy and small energy spread, which may interact with molecules or solids in that region of the Bendix mass spectrometer where the electron beam would normally pass. The product ions may subsequently be mass analysed in the normal way. Through the use of differential pumping, charge transfer and abstraction processes such as

$$A^+ + BC \longrightarrow A + BC^+$$
$$ \longrightarrow A + B^+ + C$$
$$ \longrightarrow AB^+ + C$$

may be isolated for study, at relatively low pressures and in a field-free region, so that the variation of the microscopic reaction cross-section (or rate constant) with relative velocity may be determined.

The introduction of primary ions from an external source into the normal ionization region of a mass spectrometer was first described by Lindholm.[5, 6] He coupled two mass spectrometers in tandem such that the ion beam from the one passed through the ion source of the second in a direction at right angles to the line of withdrawal of secondary ions. The same system was later used by Talroze[7, 8] who pointed out the possible discrimination of this type of system against those secondary ions produced with considerable momentum in the direction of the primary beam. Further, the geometry of any retardation lens system must ensure that no free electrons are generated at the slit edges and subsequently accelerated into the collision region. The principal disadvantage of this method is that, due to space charge considerations, a lower energy limit of a few electron volts is reached, even when using a complex lens system.[9a, b]

The earliest method of examining ion–molecule collisions by mass spectrometry was the simple one of increasing the pressure in the ionization chamber with the added variation of the repeller field to control the range of ion velocities. One of the main disadvantages of this method is that limitation of the number of types of primary ion is achieved solely by control of electron energy. The more complex the parent molecule the larger the number of fragment ions whose appearance potentials lie close together, and hence the greater the labour and possible error in correlating the appearance potentials and pressure dependencies of the various ions in order to designate them as primary, secondary, or tertiary ions. This is particularly true where the reactions of an ion of high appearance potential and low abundance are the subject of study. The more serious limitation, however, is that control of the translational energy of the reactant ions is achieved through variation of the repeller potential. Reactive collisions may occur throughout the region of acceleration from the initial point of ionization to the ion exit slit of the cage. Thus the phenomenological cross-sections determined are mean values over a certain energy range determined by the path length and ion-repeller field. A rigorous interpretation of the effect of translational energy on the reaction cross-section is therefore difficult with this approach. A comprehensive review of the data obtained using this method has been written by Lampe *et al.*[10]

Talroze and Frankevich[11] introduced the pulsed repeller technique which allowed a variable time of residence of the ions in a high pressure source before they were suddenly extracted. The technique yields rate constants for ion–molecule reactions at thermal energies and has been used extensively.[12–16] However, the translational energy range available is quite limited. An adaptation of the method was recently introduced by Futtrell and Ryan[17] whereby, by double pulsing, an initial quasi-monoenergetic bunch of ions is accelerated under a low constant field towards the ion exit slit of the cage and reactive collisions are terminated by a strong 80 V cm^{-1} pulse. With this method, the variation of microscopic reaction cross-section with relative velocity can be determined.

A novel technique developed by Cermak and Herman[18] and used by Tiernan and Futtrell[19] involves alteration of the potentials in a normal electron impact source such that primary ions are not produced in the normal chamber but only near the electron trap. The ions are then repulsed back through the normal ionization region at velocities too high for them to be extracted by the repeller field. Thus only secondary ions produced with low velocity are withdrawn and mass analysed.

A promising new approach[20] for studying ion–molecule collisions at low relative velocities is by interaction of two merging beams which have high translational energies on the laboratory scale. Thus, in the apparatus of Rothe *et al.*, a 5·1 KeV argon ion beam is deflected electrically to merge with a 5 KeV neutral argon beam which has been produced by charge transfer from 5 KeV Ar^+ beam passing through argon gas.

APPARATUS AND TECHNIQUE

Illustrated in Fig. 1 is the general layout of the additions made to the ion source end of the mass spectrometer. These are (i) a primary ion source coupled to one cross part of the Bendix ion source housing, (ii) a differential pumping system, (iii) a unit for solid sample insertion and primary ion beam collection, coupled at position X—X', (iv) various pressure measuring devices, and (v) replacement of the normal backing plate by a collector plate and guard ring for total charge collection.

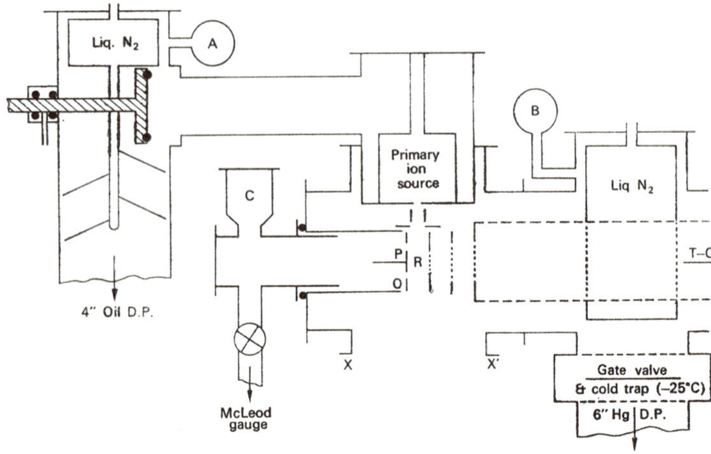

FIG. 1. Schematic diagram showing attachments to Bendix ion source. *A* and *B*, Veeco RG 75 ion gauges, *C*, Veeco BGT-75 nude ion gauge. *R*, ion–molecule and ion–solid interaction region. *P*, collector plate. *O*, guard ring.

Pressure measurement

Gas flow into the Bendix source is via a pipe which has had to be bent to face the side of the fast reaction chamber, since it was found[(3)] that, in its manufactured position pointing towards the electron beam, it gave rise to a pressure gradient across the $\frac{1}{4}$ in. gap between the backing plate and the focus grid. This was detected by sweeping a narrow primary ion beam (of width approximately 0·010 in.) across the $\frac{1}{4}$ in. gap. The secondary ion current produced by charge transfer and measured at the multiplier is proportional to the local pressure.

The actual pressure in the reaction source has generally been assessed by the ion gauge situated at position *B*. A correction factor, dependent on mass (2·2 for N_2), is necessary for this reading and is determined by recording ion peaks under static or dynamic pumping conditions.[(3)] An improved situation now exists whereby fitted on to the fast reaction chamber is the nude ion gauge *C* which can be calibrated by a McLeod gauge. These are both situated in a dead-end position and are coupled to the reaction zone by an area of approximately 2·2 cm^2 of 80% transmission fine nickel gauge. In view of this high area and 1 in. wide bore tubing the ion gauge pumping is considered negligible. The McLeod gauge has to be operated under such conditions that the Ishii–Nakayama effect is minimized.[21-23] The primary source pressure is monitored by ion gauge A.

In cases where the total cross-section for ionization by electron impact at a particular energy is known, an internal method of pressure calculation may be made. This involves measurement of the total positive ion current i_+ collected on the collector plate and use of the relation $i_+ = Q_e i_e dp$, where i_e is the electron trap current, d is the path length (in our case 10 mm), and Q_e is the total ionization cross-section for electrons at the specified energy. Recently Harrison *et al.*[24] have given some values for Q_e and compared them with those of Otvos and Stevenson[25] and Lampe *et al.*[26]

The degree of differential pumping may be gauged from the data that with the slit sizes in electrodes C and D set at 9×1 mm and $3 \times 0{\cdot}5$ mm respectively, introduction of nitrogen into the primary source leads to a value 135 for the ratio of ion gauge A/ion gauge B readings, whereas introduction of gas into the Bendix source leads to a value of 450 for the ratio of ion gauge B/ion gauge A. Introduction of gas via the neutral beam inlet leads to a drastic reduction of these values.

External ion source

The ion source arrangements are shown in greater detail in Fig. 2. The primary ion beam of singly charged, monatomic or diatomic, ions is generated by electron energy control and may be either continuous or pulsed. Electrodes A–F are accelerating, focusing or retardation electrodes and the ion beam A^+ interacts with gas BC or solid sample in the region R. Secondary ions produced there are either collected *in toto* on the collector plate P and a total reaction cross-section or yield determined, or are withdrawn into the flight tube by a -280 V pulse on the ion extraction grid and mass analysed. In the latter case, ion noise is more troublesome when continuous primary beams are used. This is due partly to the continuous withdrawal of secondary ions through a combination of thermal diffusion and some field penetration from the energy grids. It is also partly due to deflection by the focus pulse (-280 V) of the primary ion beam itself,

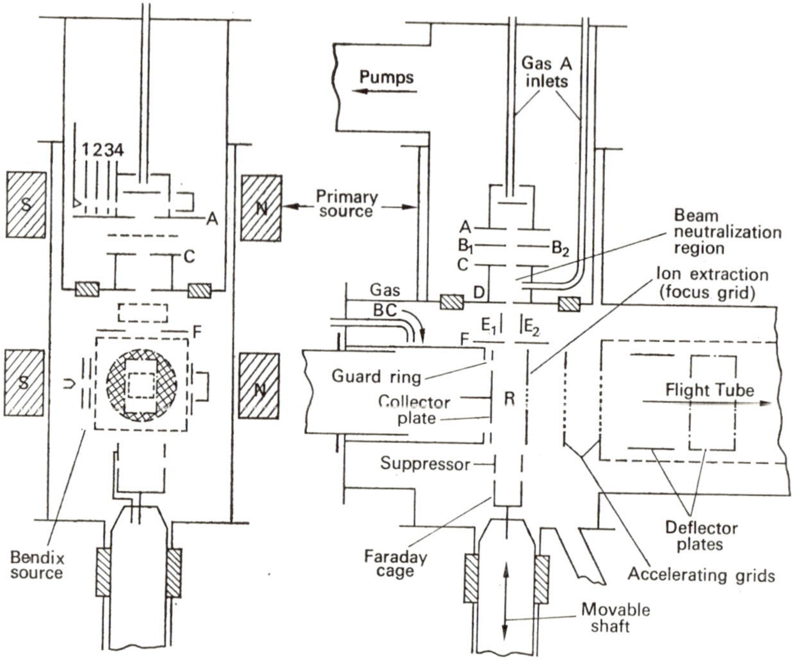

FIG. 2. Schematic of Bendix and auxiliary ion sources; two cross-sections at right angles. Bendix source magnets normally removed.

in a certain energy range, on to the extraction aperture in the ion focus grid. The aperture size has been reduced from the original 19 mm diameter to 7×7 mm square, principally in order to increase the path length for atom transfer reactions, and also to improve the collection efficiency of secondary ions produced by charge transfer. A subsidiary source of ion noise caused by ion gauge *B* was eliminated by insertion of a 90° metal elbow into the gas flow line.

When using the continuous mode of operation, improved intensity of the primary beam may be achieved by using the electrode slits 2 and 3 as focusing plates. Previously,[(3)] the arrangement was that at any nominal electron voltage in the range 5–65 V, the plates 2, 3 were tied electrically to the cage and plate 4, with plate 1 at −65 V with respect to the filament. A decided improvement, particularly at low electron energies, is achieved through biasing separately the electrodes 2 and 3 to positive voltages with respect to cage. The transmission efficiency defined as trap current/cage plus grid 1 current is then $>10\%$ at a nominal electron energy of 10 eV, whereas previously it would be $<1\%$. One field of interest to which the above improvement is particularly relevant is in the variation of cross-section of charge transfer reactions as a function of the degree of excitation of the primary beam, the latter being controlled simply by electron energy control.

In the model S-14-105 ion source, as originally supplied, the central axis of the electron beam was distant 0·085 in from the backing plate and 0·165 in. from the trapping (focus) grid. For our work, involving the use of an externally created beam of ions instead of electrons as the ionizing agent, the insertion of Faraday cages and suitable solid surfaces for sputtering into the gap between the backing plate and the trapping grid was greatly facilitated through the use of electron and ion beams which crossed in a central plane. Wiley and McLaren[27] suggested that decrease of the distance between the electron beam and the trapping grid should decrease the resolution. We tested for this effect experimentally through the use of a 3 mm wide slit at F (Fig. 2) and scanning an ion beam of constant intensity across the central two-thirds of the gap between the backing plate and trapping grid, using the E_1–E_2 deflector plates. With both mercury and cumene at constant pressure in the Bendix source, we noted no perceptible change in either mass resolution or sensitivity. There was, however, a shift in the time of flight of the secondary ions (see later under Atom Transfer, p. 193).

Unit for primary ion beam collection and solid sample insertion

The sample probe system (Fig. 3) is attached to the Bendix cross-port at position X–X' in Fig 1, directly opposite to the auxiliary ion source. It provides basically for (a) primary beam current measurement, and (b) sample insertion for sputtering studies. Collection of the primary ions may be made either in the Faraday cage shown in Fig. 3 when the cage is moved inwards to intercept the ion beam or, as is preferable, in a deep Faraday cage (15×5×7 mm) and suppressor electrode mounted on a duplicate main shaft. The latter procedure has a distinct advantage in that it enables the cage to be manipulated along the vertical axis of the primary ion beam to positions between the backing plate and extraction (focus) grid. Thus the focusing of the primary ion beam may be investigated as a function of the cage position. The focusing has been adjudged to be satisfactory for charge transfer studies down to 75 eV. Space charge defocusing is such

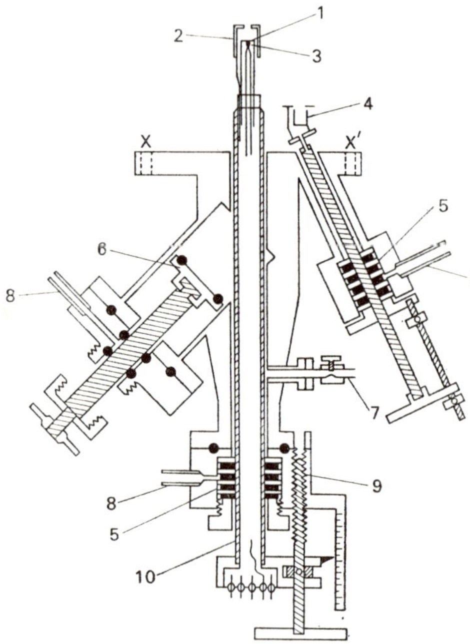

Fig. 3. Unit for solid sample insertion and primary beam collection, coupled at position $X-X'$ in Fig. 1. 1, filament plus sample. 2, repeller box electrode. 3, Pallador thermocouple. 4, auxiliary ion collector and electron suppressor, shown in withdrawn position. 5, alternate steel and polytetrafluoroethylene washers compressed on to stainless steel shaft. 6, vacuum lock in open position. 7, initial pump-down line. 8, auxiliary pumping lines between vacuum seals. 9, guide pin and target position screw adjustment. 10, hollow non-magnetic movable shaft.

that, at primary energies of 5 eV, only 80% of the beam is collected by the Faraday cage placed just beyond the extraction zone, i.e. 25 mm beyond slit *F*. Lindholm and Gustaffson,[9a, b] using a sophisticated multi-electrode system, have decelerated ion beams down to 2 eV. The above limitation does not apply in our technique for studying atom transfer reactions in the energy range 2–60 eV, where both primary and secondary beams are estimated by the same method. Here it can be shown from momentum considerations that any losses will be comparable for both primary and secondary ions provided the masses of the target gas and atom transferred are small compared with the mass of the reactant ion.

The main shaft insert was originally[1] sealed by greased O-rings. This method has been discarded in favour of one consisting of compressed alternate teflon–stainless steel washers which has proved highly reliable and successful. The time for sample charge has been reduced from 10 to 3 min, with the flight tube pressure at no stage rising above 10^{-6} torr. The appearance of mass 74 ($SiMe_2O^+$) has been considerably reduced.

Primary beam currents are measured either by a Vibron electrometer Model 62A (sensitivity range 10^{-16}–10^{-7} A) or by one of the Bendix electrometers with the meter circuit reversed. Typical intensities of 20 μA cm^{-2} are achieved with the Bendix source background pressure rising approximately 3×10^{-7} torr. The latter rises disproportionately with any further increase in beam intensity.

The optimum position of the solid surface which is to be sputtered is 9·5 mm below the flight tube axis and the positioning is fairly critical for optimum results ($\pm$0·5 mm). The main difficulties which have been overcome are:

(a) In the case of continuous bombardment, ion noise was not constant throughout the cycle but a function of primary ion mass and position of the surface.
(b) The chemical nature of the surface affected stability. Thus for non-conductors charge-up resulted in a pulsating spectrum.
(c) Adsorption layers may be very significant at room temperature when pulsed beams are used.

The oscilloscope display of the mass spectrum is only used for optimizing the experimental parameters prior to any runs. The actual abundances of secondary ions are measured with the two-channel analogue output system at relatively slow scan (i.e. high sensitivity).

Timing of pulsed beams

The requirements for pulsing the primary source are illustrated in Fig. 4. In the case of ion–molecule gas phase reactions, a 10 V pulse of 50 nsec duration is applied to grids 2 and 3 of the primary source. The resultant primary ions are focused through the reaction zone where they interact with neutral molecules, after which a pulse is supplied to the focus grid to draw out the result-

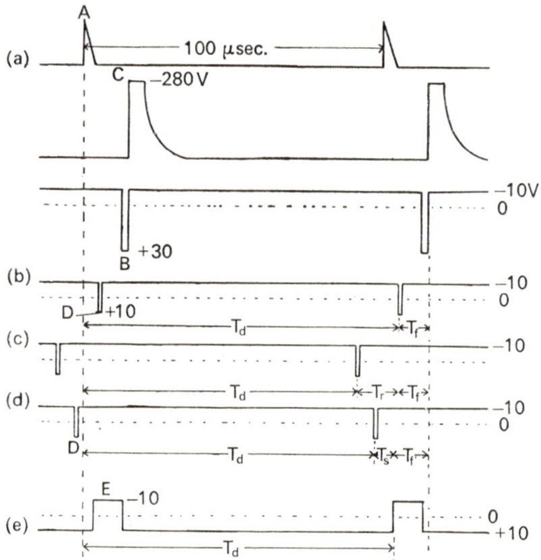

FIG. 4. Timing of pulses. (a) Normal Bendix cycle involving electron impact. *A*, Bendix master synchronizing pulse. *B*, Bendix electron pulse of 0·25 μsec duration. *C*, Bendix ion withdrawal and focusing pulse. (b) Pulsed primary source operation where the primary ions arrive in the reaction zone immediately before the extraction pulse is applied. *D*, electron pulse supplied to primary source grids 2, 3 which are biased −10 V d.c. with respect to filament. T_f, flight time of primary ions from primary source to reaction zone. T_d, delay triggered by the master pulse of the previous cycle. (c) Pulsed primary source operation where the primary ions arrive at a time T_r prior to application of the extraction pulse. T_r, residence time of secondary ions. (d) Pulsed primary source operation when sputtering at a duty cycle in the range of 0·05–5%. T_f', flight time of primary ions from primary source to target. T_s, flight time of ions sputtered from the target to the extraction zone. (e) Pulsed primary source operation when sputtering at 90% duty cycle. Ion bombardment is suppressed by pulse *E* whilst mass analysis is made. Grids 2, 3 are biased +10 V d.c. with respect to filament.

ant secondary ions for mass analysis. The delay between the electron pulse and the draw-out pulse may be adjusted so that the latter is applied either as soon as the primary ions have passed through or after a selected interval of time. This variable delay permits the study of the diffusion of the secondary ions, the decomposition of metastable ions and, at slightly increased pressures, of further ion–molecule (e.g. condensation) reactions at essentially thermal energies. The degree to which linear momentum is transferred from the primary to the secondary ions is strongly dependent on the type of reaction process.

Two pulsing methods have been employed in studies of the sputtering of solids. These are (a) the use of 80–90% duty cycle pseudo-continuous bombarding beams, and (b) the use of pulsed beams of 0·05–5 μsec duration. The advantages of the method (a) are (i) that the surface is kept relatively clean from adsorbed gases, and (ii) that a relatively noise-free mass spectrum is obtained through suppression of the primary beam and secondary ion generation in the short interval of time just before and during the draw-out pulse. The timing here is such that the primary beam is suppressed at such a moment that when pulse C comes on, no A^+ ions are sufficiently far along their flight path as to be drawn into the transmission grid of the ion focus plate to create ion noise. The percentage duty cycle is limited by the chosen mass range for analysis. This is because the lighter mass ions produced at the leading end of a second pulse (approximately 100 μsec later) will arrive unfocused at the multiplier at the same instant as the focused heavy mass ions produced at the tail end of the first pulse. Lower limits to the usable mass range are m/e 1–60 at 90% duty cycle and 1–200 at 80% duty cycle.

One of the pitfalls to be avoided when using method (b) is that unless the pulse is sufficiently long for the particular chemical system being studied, a non-representative sample of the ions sputtered will be withdrawn. The shorter the mass range of interest, the shorter the pulse that may be used; such that at the instant pulse C is applied, sufficient overlap of the faster light ions with the heavy ions will have occurred directly opposite to the extraction grid. In addition, under both methods of operation, a further factor contributes to mass discrimination, namely that the number density of sputtered ions opposite the extraction grid varies as $(m/e)^{1/2}$. In theory, one should be able to eliminate these restrictions (i) by arranging that the solid surface is at or just above the level of the lower side of the extraction grid, and (ii) by using very short duration primary pulses. In practice, we have found that this method leads to decreased sensitivity and resolution, due no doubt to distortion of the focusing field lines between the ion focus grid and the collector plate by the solid sample insert. Thus for each particular chemical system being studied, consideration of the three factors of sensitivity, resolution and discrimination demands that a compromise be effected involving (a) the size of the extraction hole in the focus grid, (b) the primary pulse width, and (c) the distance between the target surface and the lower side of the extraction hole.

Mass analysis of the primary ion beam

Provided the translational energy E_p of the primary ion beam is low enough, the beam may be withdrawn by the -280 V extraction pulse and mass analysed in the normal way. At present with a 7 mm square hole in the extraction (focus) grid, the upper limit for E_p is 60 eV, whereas previously with a 19 mm diameter hole the limit was about 150 eV. As an aid for this way of analysis, the vertical deflector plates are arranged to be effective before the horizontal deflector plates in the flight tube.

In experiments on the ion bombardment of either gases or solids, the Bendix source magnets are removed. This is to reduce any discriminating effects of the crossed electric and magnetic fields on (a) the primary ion beam at low energies, and (b) the trajectories of the secondary ions (which may have energy half-widths of several electron volts). The primary source magnetic field is approximately 100 gauss, but the fringing field at the interaction zone is not considered excessive.

Analysis of the target and background gases in the collision chamber is made using the Bendix electron beam with the corresponding source magnets replaced temporarily for increased sensitivity.

Sensitivity and linearity of response

The sensitivity of the basic time-of-flight instrument is such that, at a trap current of 0·2 μA and 70 eV electron energy, a peak corresponding to a single ion per cycle is equivalent to a partial source pressure of argon of 5×10^{-8} torr. With the multiplier gain at its maximum, such a peak yields an electrometer output of 1 mμA, indicating a gain of 6×10^5. On maximum sensitivity, a reading of 2×10^{-12} A (or 1 mm deflection on a chart recorder) corresponds to a partial pressure in the source of 2×10^{-12} torr of argon. The largest signal which can be recorded is $\simeq 70$ mμA when the linear response of the multiplier ceases.

With oscilloscope presentation, the usable range is 50 : 1 and 20 : 1 at low and maximum multiplier gain respectively. Thus in the latter case a single ion peak yields an output peak of 0·1 V, and at 2·0 V output non-linearity of response of the IFI wide band amplifier occurs.

The linearity of output signal with source pressure (with no baffling) fails in the region of 2×10^{-5} torr due to collisional losses down the flight tube. With the total charge collection method, involving no mass analysis, the pressure limit is raised to the region of 8×10^{-5} torr. Condensation reactions are thus best studied by a suitable combination of increase of pressure with an increase of time of residence (T_r in Fig. 4c).

The effect of reducing the voltage (-2800 V) on the central ion accelerating grid of the Bendix source was tested in the normal pulsed mode of operation of the instrument. The sensitivity was improved by a factor of 2·0 at 400 V, 2·5 at 700 V, and 2·0 at 1000 V.

Determination of reaction cross-section

All reaction cross-sections have been determined under conditions such that the incident particle density is essentially constant throughout the reaction path length, i.e. there is negligible loss. Hence the general equation

$$I_S = 0{\cdot}304\, I_P d P Q_{R(v)} f \tag{1}$$

may be used, where I_P and I_s are the primary and secondary currents, d is the reaction path length (in cm), P is the reactant gas pressure (in torr), and $Q_{R(v)}$ is the cross-section (in Å^2 units) corresponding to the selected relative velocity v of collision; f is a calibration factor, the value of which depends on which method of collection and measurement of I_S is employed. If mass analysis is made, then only relative cross-sections may be determined, owing to the uncertainty of such factors as the multiplier gain and the transmission efficiency of secondary ions from source to multiplier. In this mode of operation, calibration of the instrument to yield absolute cross-sections may be made by use of a standard reaction. Thus two values of f have been determined;[3] (a) $f = 3 \times 10^5$ using $Q_R = 25$ Å^2 for the Ar^+/Ar charge transfer process at 400 eV collisional energy[28] and (b) $f = 2{\cdot}5 \times 10^5$ using electron impact ionization data[29] for nitrogen.

We equate the factor f to unity in two particular cases, namely (i) in atom transfer cross-section measurements where the primary and secondary ions are assumed to be collected, mass analysed, and measured with equal efficiency, and (ii) in the determination of total charge transfer cross-section. The technique employed in case (ii) is to apply a potential difference between the focus grid and the backing plate so that slow-moving secondary ions are attracted uniformly to both a collector plate P (10×23 mm) and its surrounding guard ring O which is kept at the same potential (Fig. 2). The magnitude of the collection potential is sufficiently small to have negligible effect on the trajectory of the primary ion beam. The current to the collector plate I'_S is meas-

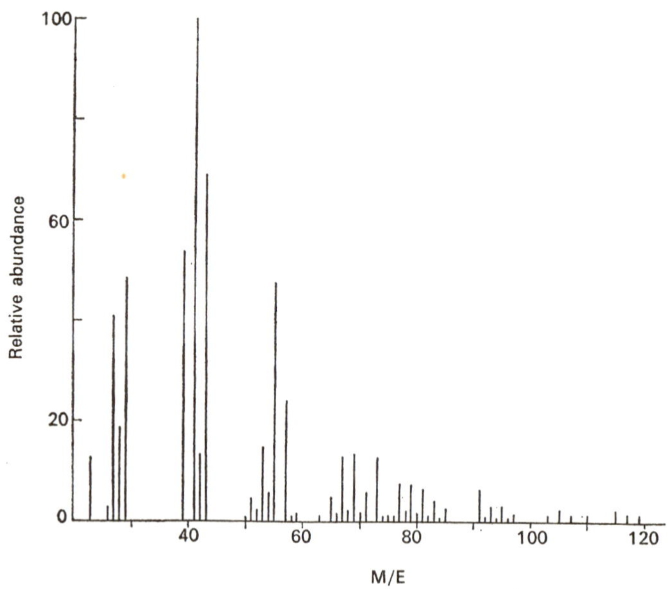

FIG. 5. A typical mass spectrum of the positive ions sputtered from a stainless steel collector or backing plate at 20°C.

ured as a function of the collection potential over the range 0–20 V. This procedure is repeated under identical conditions with the reactant gas supply cut off. The difference $\Delta I'_S$ between these sets of values is plotted against collection potential. A typical collection curve would rise steeply to an almost horizontal region which on extrapolation back to the $\Delta I'_S$ axis yields a limiting value I_S, the ion current corresponding to secondary ions with negligible momentum. Application of eqn. (1) then yields $Q_{R(v)}$, and the corresponding rate constant is given by $k_{(v)} = vQ_{R(v)}$.

Cleanliness of the ion source is important. The collection potentials required to reach plateaux in the various collection curves were initially high ($\sim$ 20 V) suggesting that secondary ions with excessive kinetic energies were being formed. The stainless steel plates readily adsorb carbonaceous matter, as shown in Fig. 5, and it is probable that surface charges interfered. An Aquadag coating[30] of all the grids, plates and Faraday cages in the ion source region drastically reduced the necessary collection potential to approximately a volt. An alternative method is to operate the source region at an elevated temperature.

Multiplier discrimination

One of the major uncertainties in cross-section measurements with instruments which have electron multiplier detection systems is the response of the first dynode (cathode) to the type of ion being collected. A general observation applicable to the Bendix TOF instrument is that a recorded mass spectrum is reasonably similar to those which have been obtained using magnetic deflection instruments and are quoted in API tables. Differences are greater the lower the fragment ion mass. In addition to the cathode effect, the Bendix TOF and various magnetic deflection instruments which employ fine slits may well have different powers of discrimination against ions possessing kinetic energy. The maximum divergence from the TOF axis of ions possessing kinetic energy E_d as a result of a dissociation process is given by $L(E_d/E)^{1/2}$, where L is the flight path (100 cm) and E is the flight energy (2800 eV). If, for example, $E_d = 1$ eV, a maximum divergence of 1·3 cm is obtained which would be too large for total ion collection on to the multiplier cathode, unless there is extra focusing. In addition, for those ions collected, two factors contribute to the peak width; firstly the effect of initial kinetic energy on the distribution of flight times, and secondly the variation of electron transit times across the first dynode on account of the variation of the position of ion impact.

Futtrell[31, 32] has examined the first dynode effect in some detail and finds typical discrimination factors to be

1·82 for water	H_3O^+/H_2O^+
1·51 for ethylene	$C_2H_5^+/C_2H_3^+$
1·00 for carbon monoxide	COH^+/CO^+

With hydrocarbons he found the gain was a function of ion *m/e* and history of the multiplier. He compared the mass spectra of hydrocarbons in a mass spectrometer which had alternately an electrometer and a multiplier output system. A smooth curve was obtained for the electrometer pattern coefficient divided by the multiplier pattern coefficient from a value of 0·6 at *m/e* 12 to 2·0 at *m/e* 52.

Some systematic investigations on other types of multiplier have been made by other workers[34, 35] as a function of such parameters as mass and chemical structure of the ion, charge, velocity and angle of incidence. The main conclusions are that, in the range of 1–5 keV energy, the gain is approximately a linear function of energy and that at equal energy the gain decreases with increase in mass. C. la Lau[35] studied the response of a copper–beryllium type multiplier as fitted onto an AEI MS 9 mass spectrometer. He found (a) the response of three heavy ions $C_{24}H_{48}^+$, $C_{26}H_{26}^+$ and $C_6H_3Br_3^+$ at 9 keV energy was 1·4, 1·3 and 0·65 respectively, relative to the $C_3H_5^+$ ion taken as standard, (b) the relative yield for hydrocarbon ions increased smoothly from $C_2H_5^+$ (1·04) to $C_{13}H_{27}^+$ (1·30). He showed that, provided the individual response curves of each atom as a function of velocity is known, then application of the additivity principle allows one to calculate the response of any complex ion at a chosen energy to within 10% accuracy.

There can be little doubt that the multiplier response is also a function of its history. This has been investigated by Futtrell and Tiernan[32] and indirect support is given by our observation that the noise level of the Bendix multiplier gradually increases with time. With all ion sources and ion gauges switched off, a freshly cleaned multiplier gives undetectable ion noise ($< 2\times10^{-13}$ A total signal output) at maximum gain at 10^{-7} torr pressure. This level rises with pressure to 25 pA at 2×10^{-5} torr and 0·2 nA at 5×10^{-5} torr. These values increase with age and an obviously dirty multiplier which gives rise to an unstable mass spectrum has a noise level in the region of 0·1 nA at 10^{-7} torr.

RESULTS

To illustrate the type of data that we have obtained with the present apparatus, a few representative results will be described. Molecules may be converted to positive ions by interaction with either ions or neutral particles. The main reactions which have been of interest to us may be classified as:

(a) Symmetrical charge transfer, e.g. $Ar^+ + Ar \rightarrow Ar + Ar^+$.
(b) Unsymmetrical charge transfer, e.g. $Ar^+ + N_2 \rightarrow Ar + N_2^+$.
(c) Dissociative charge transfer, e.g. $Kr^+ + CH_4 \rightarrow Kr + CH_2^+ + H_2$.
(d) Atom transfer, e.g. $Ar^+ + D_2 \rightarrow ArD^+ + D$.
(e) Neutral-impact ionization, e.g. $He + CH_4 \rightarrow He + CH_3^+ + H + e$.

Ion-impact ionization is not significant compared with one of the charge transfer processes unless the energy of collision is a few keV or greater.

The cross-sections for all the above processes vary with translational energy and the characteristics of these variations may provide information about the mechanisms of the processes and of the energy states of the particles involved.

Charge transfer

The four simple charge transfer reactions between argon and nitrogen ions with both argon and nitrogen have been examined[36] and evidence presented for the occurrence of non-resonant symmetrical and unsymmetrical processes. In these processes, there is an energy defect $\Delta\varepsilon$ which is equal to the ionization potential of the reactant particle minus the recombination energy of the reactant ion. The reaction cross-section for these processes is expected to pass through a maximum at a specific translational energy E_m. The near-adiabatic hypothesis of Massey and

Burhop[37] relates $|\Delta\varepsilon|$ to v_m, the relative velocity of impact corresponding to maximum cross-section

$$|\Delta\varepsilon| \simeq hv_m/a, \tag{2}$$

where h is Planck's constant and a is a length of atomic dimensions. For a reaction involving single electron transfer to a singly charged ion, this implies the following expression for E_m:

$$E_m \simeq 3{\cdot}08\, m_1(\Delta\varepsilon)^2 a^2,$$

where m_1 is the mass of the reactant ion in atomic units and E_m and a are expressed in eV and Å units respectively. A mean value of the parameter a of 7 Å with a standard deviation of 30% has been calculated by Hasted[38] from an examination of more than 100 systems assuming an equality in eqn. (2). The absence or presence of a maximum in the curve relating cross-section to translational energy in a charge transfer system indicates whether the reaction is non-resonant or resonant. In the latter case, the cross-section is expected to decrease with relative velocity according to the equation[39, 40]

$$Q^{1/2}_{R(v)} = c - b \ln v, \tag{3}$$

where v is the relative velocity and c and b are constants.

Figure 6 illustrates the data obtained for the $N_2^+ + Ar$ and $Ar^+ + N_2$ charge transfer systems. It was deduced by application

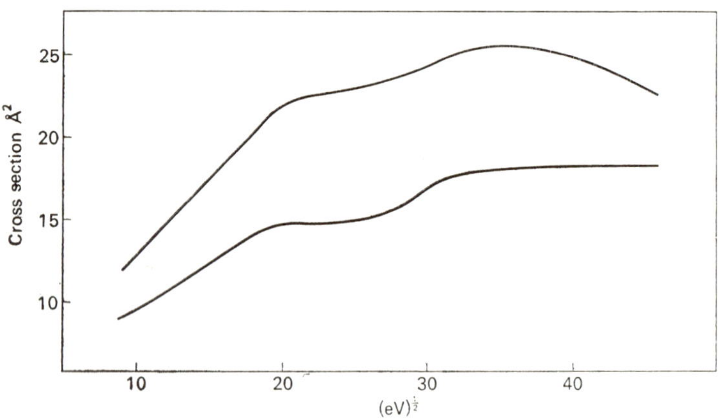

FIG. 6. Unsymmetrical charge transfer reactions; cross-section plotted against square root of translational energy. *Upper curve:* $N_2^+ + Ar$; *lower curve:* $Ar^+ + N_2$.

of the above theory that the two processes occurring in the $Ar^+ + N_2$ system were:

$$Ar^+ (^2P_{3/2}, 15{\cdot}76 \text{ eV}) + N_2 \longrightarrow Ar + N_2^+ (X^2\Sigma_g^+, 15{\cdot}58 \text{ eV}); \Delta\varepsilon = +0{\cdot}18 \text{ eV},$$

$$Ar^+ (^2P_{1/2}, 15{\cdot}94 \text{ eV}) + N_2 \longrightarrow Ar + N_2^+ (X^2\Sigma_g^+, 15{\cdot}58 \text{ eV}); \Delta\varepsilon = +0{\cdot}36 \text{ eV},$$

and the two reactions in the $N_2^+ + Ar$ system were the exact reverse. The near-adiabatic parameter *a* had a value in the range 10–12·4 Å in each case. The assumption that N_2^+ is formed in the ground state is justified since the lowest excited states, $A^2\Pi_u$, 16·7 eV and $B^2\Sigma_u^+$ 18·76 eV, would lead us to expect E_m values greater than 4000 eV. An alternative interpretation of the $N_2^+ + Ar$ system assumes that the nitrogen product ion is formed in the first vibrational level at 0·287 eV. In this case, the reactions are:

$$N_2^+ (15{\cdot}58 \text{ eV}) + Ar \longrightarrow N_2(0{\cdot}29) + Ar^+(15{\cdot}76); \Delta\varepsilon = -0{\cdot}47 \text{ eV},$$

$$N_2^+ (15{\cdot}58) + Ar \longrightarrow N_2(0{\cdot}29) + Ar^+(15{\cdot}94); \Delta\varepsilon = -0{\cdot}65 \text{ eV}.$$

The decreased recombination energy of 15·29 eV leads to near-adiabatic parameters of 6·5 and 4·7 Å corresponding to E_m = 1450 eV and E_m = 400 eV respectively.

The primary ion beams in the above studies were produced by electron impact at energies just below the second ionization potential in order to achieve high intensities. Hence the statistical weights of the $^2P_{3/2}$ and $^2P_{1/2}$ in the Ar^+ beam would be in the ratio 2 : 1. However, the type and degree of excitation in the Ar^+ and N_2^+ beams are unknown. Long-lived excited Ar^+ ions have been detected at appearance potentials of approximately 33 V by Hagstrum[(41)] and 43 V by McGowan and Kerwin.[(42)] That the method of ion beam generation can have a decided effect on cross-section versus translational energy curves was strikingly shown by Fogel *et al.*[(43)] He examined the $Li^+ + Ar \longrightarrow Li^- + Ar^{++}$ reaction using Li^+ ions produced by both electron impact and surface ionization, the latter process yielding exclusively ground state ions. Since then, studies by Utterback and Amme[(44)] have shown a decrease in N_2^+/N_2 charge transfer cross section of 15% as the electron energy used to produce the primary N_2^+ ions was increased from 17 to 23 eV. For the same system McGowan and Kerwin[(45)] observe a 16% decrease as the electron energy goes from 16 to 23 eV. The curves are smooth and it is not possible to distinguish the effects of the $A^2\Pi_u$ (16·7 eV) and $B^2\Sigma_u^+$ (18·75 eV) excited states of N_2^+.

At thermal energies, McGowan *et al.*[46] have demonstrated near-resonant charge transfer reactions in the $Ar^+ + N_2$ and $N_2^+ + Ar$ systems for different vibrationally (N_2^+) and electronically (Ar^+) excited levels. This has been achieved by an analysis of the ionization efficiency curves of the two gases separately and in mixtures.

Dissociative charge transfer

One approach to the study of dissociative charge transfer reactions is to correlate the energy requirements of an endothemic process with its translational energy threshold. If $\Delta\varepsilon$ (equal to recombination energy minus appearance potential) is negative the reaction path is endothermic and a threshold energy may be predicted for the process from conservation laws. Comparison may then be made with the experimentally determined threshold energy, which is obtained by extrapolating cross-section versus translational energy curves to zero cross-section.

The data obtained for the $Kr^+ + CH_4$ system[36] is shown in Fig. 7. No thresholds are predicted for the reactions leading to the formation of CH_4^+ (a.p. = 13·1 V) and CH_3^+ (a.p. = 14·4 V) despite the fact that the i.p. of Kr is 14·00 V. This is because the krypton ion beam will contain approximately 33% of the upper doublet state $^2P_{1/2}$ (r.e. = 14·67 eV), so even the CH_3^+ may be formed exothermally. The threshold condition for any endothermic process is

$$E_T = \Delta\varepsilon(m_{Kr} + m_{CH_4})/m_{CH_4}, \tag{4}$$

where m is the mass of the particle and E_T is the translational energy threshold for the process whose endothermicity is $\Delta\varepsilon$. The data in Fig. 7 indicates the existence of thresholds in the region of 10 ± 10 and 50 ± 10 eV for the processes leading to CH_2^+ and CH^+ formation respectively. The three reactions to be considered are:

(a) $Kr^+ + CH_4 \longrightarrow Kr + CH_2^+ + H_2$; $E_T = 5·8$ eV (Kr^+, $^2P_{1/2}$) or 10·0 eV (Kr^+, $^2P_{3/2}$).

(b) $Kr^+ + CH_4 \longrightarrow Kr + CH_2^+ + 2\,H$; $E_T = 34$ or 38 eV.

(c) $Kr^+ + CH_4 \longrightarrow Kr + CH^+ + (3\,H)$; $E_T = 52$ or 56 eV.

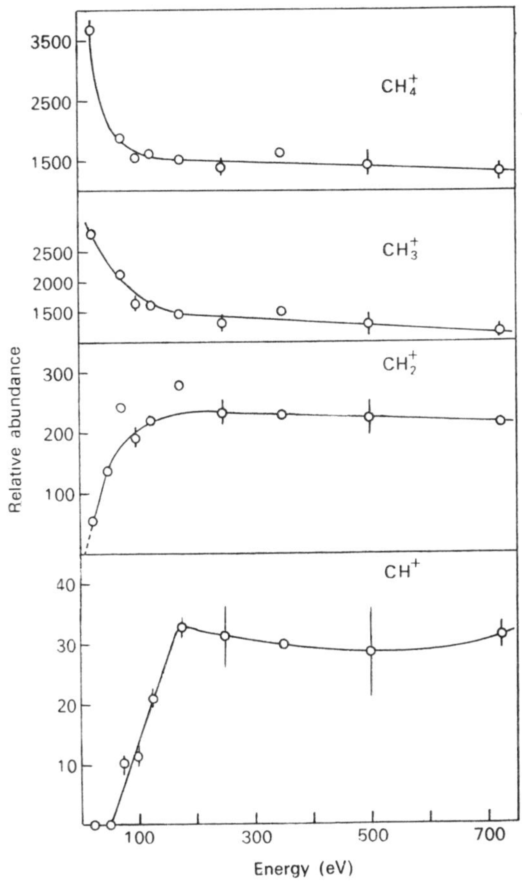

FIG. 7. Dissociative charge transfer reactions in the system $Kr^+ + CH_4$; cross-sections as functions of energy.

The values of E_T are the calculated thresholds using both states of the Kr^+ ion and assuming appearance potentials of 15·6 and 20·1 V for CH_2^+ and 23·0 V for CH^+ production from CH_4 by electron impact. The evidence is that reaction (a) rather than (b) occurs in the threshold region, but the possible occurrence of reaction (b) above its predicted threshold at 34–38 eV cannot be excluded.

Talroze[8] examined in detail the process

$$Xe^+ + C_2H_4 \longrightarrow Xe + CH_2^+ + CH_2; \ \Delta\varepsilon = \begin{cases} 7{\cdot}17 \text{ eV } (Xe^+, {}^2P_{3/2}) \\ 5{\cdot}86 \text{ eV } (Xe^+, {}^2P_{1/2}) \end{cases}$$

for which the calculated energy thresholds are 33·3 and 40·7 eV respectively. The experimental threshold was in the region of 35 eV in good agreement. The excitation function showed an increase immediately after the threshold considerably steeper than that implied by the Massey near-adiabatic hypothesis. This conclusion applies equally to our data in Fig. 7, and is not surprising since the observed cross-section is a summation $Q = \sum_{ij} F_{ij}Q_{ij}$, where the number of accessible energy states in the complex molecules methane and ethylene will be large. F_{ij} is the fraction of the ion beam which is in state i and reacts to yield product ion in state j.

Maier[47–49] has found other examples of reaction thresholds and has shown how, provided that there is no excitation in the products and no kinetic energy release in the centre of mass sys-

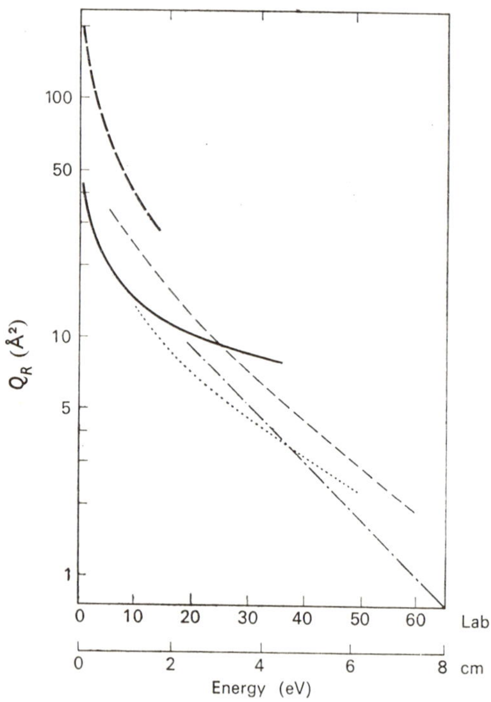

FIG. 8. Cross-section of $N_2^+ + D_2 \longrightarrow N_2D^+ + D$ as a function of energy. ———— orbiting theory.[52] — — — — — — Giese and Maier.[53] — — — — Homer *et al.*[54] .... Turner *et al.*[56] —·—·—·—·— Lacmann and Henglein.[57]

tem, threshold energy determinations may lead to bond dissociation energy values. An analysis in terms of phase space has been developed by Light.[50]

At translational energies well above their respective thresholds, the abundancies of all fragment ions continue to be dependent on the energy of collision, contrary to the conclusion of Talroze.[8] In the case of methane[2, 51] the dependence is relatively weak and the mass spectrum is approximately that to be expected on the resonance principle whereby the ion produced in maximum abundance is the one for which the internal energy defect $\Delta\varepsilon$ is least. However, in the case of rare gas ion impact on ethylene and propane, the dependence is complex and the cross-sections for a few fragment ions may increase or decrease several fold in the energy range 100–2000 eV. Due to the multiplicity of energy states and potential curve crossing, the resonance principle certainly cannot be applied to these systems. It has been suggested[4] that the fine structure observed in the cross-section versus translational energy curves can give some indication of the breakdown mechanisms.

Atom transfer

Most of the information on reactions of the type $A^+ + BC \longrightarrow AB^+ + C$ has been derived from studies of the pressure and repeller voltage dependencies of the various ions obtained from a mass spectrometer source to which a mixture of the gases A and BC has been supplied. The limitations and disadvantages of this method have been discussed earlier, in particular the translational energy range over which the reaction cross-section may be determined.

The earliest proposal[52] for the mechanism of such reactions was that there is initial formation of an orbiting complex between A^+ and BC followed by transfer of atom B. An orbiting complex is formed if the attractive polarization force between the ion and the molecule is equal to or greater than the centrifugal force which tends to dissociate the complex. The maximum impact parameter (the impact parameter is the perpendicular distance between one particle and the initial line of relative motion of the other) which can lead to the formation of an orbiting complex is taken as the radius of the reaction cross-section. Thus the reaction cross section Q_v for an interaction between an ion and a molecule with an initial relative velocity of v is deduced to be

$$Q_v = \frac{2\pi e}{v}\left(\frac{\alpha}{\mu}\right)^{1/2}, \tag{5}$$

where e is the charge of the ion, α the polarizability of the molecule and μ the reduced mass of the system.

A few years ago several workers[(53-56)] examined the above relationship using beam techniques with which the velocity range was increased by orders of magnitude over previous studies. The method we have used is to decelerate the primary beam to its final energy at slit F (Fig. 2) and then withdraw the product ions through the 7 mm square hole in the trapping grid. The distance from slit F to the top of the extraction hole, 18 mm, is taken to be the reaction path length, neglecting the small indeterminate errors involved at each end of the path length where the ions are rapidly decelerated and accelerated. Provided its energy is $\leqslant$ 60 eV, the primary ion beam may be withdrawn into the flight tube by application of the -280 V ion focus pulse. The vertical deflector potential has to be re-set from what it is in the case of thermal ion production in the Bendix source. This is due to the momentum of the primary beam.

Scattering of the beam will be minimized, and hence collection of both unreacted and product ions facilitated, in cases where a heavy ion bombards a light target gas and the product ion retains a high proportion of the primary ion momentum. The system has the advantage over the repeller dependence type of study of covering a wider energy range (up to 60 eV with the present 7 mm square hole) and securing a relatively narrow velocity distribution within the reaction zone which is field free. The intensities of both A^+ and AB^+ are thus recorded simultaneously. Similarity of mass of A^+ and AB^+ also has the advantages of similarity of flight path, ion velocity and multiplier gain. Thus, assuming equal collection efficiency of reactant and product ions, and operating under such conditions that the primary ion beam attenuation is low, one can calculate the reaction cross-section Q_R by eqn. (1) where the factor f is unity.

In practice, one of the difficulties encountered is a tailing on the high mass side of the primary peak which reduces the resolution for atom abstraction studies to about 40. This probably arises from the broadening of the primary beam as its energy E_p decreases. Complete collection of i_p in a Faraday cage placed at the end of the abstraction zone has been shown (by means of a movable Faraday cage) to cease at $E_p = 40$ eV. In other words, the primary beam has broadened from 0·5 mm to $\geqslant$ 5 mm after travelling at 75 eV energy for 18 mm beyond the *F* slit. The percentage collection of primary beam at the same position at $E_p = 5$ eV has been shown to be 90%. This broadening of the beam is to some extent compensated for by the space focusing action of the Bendix ion source.[27] We have made the observation that using a high energy primary beam and the E_1–E_2 deflector plates one can

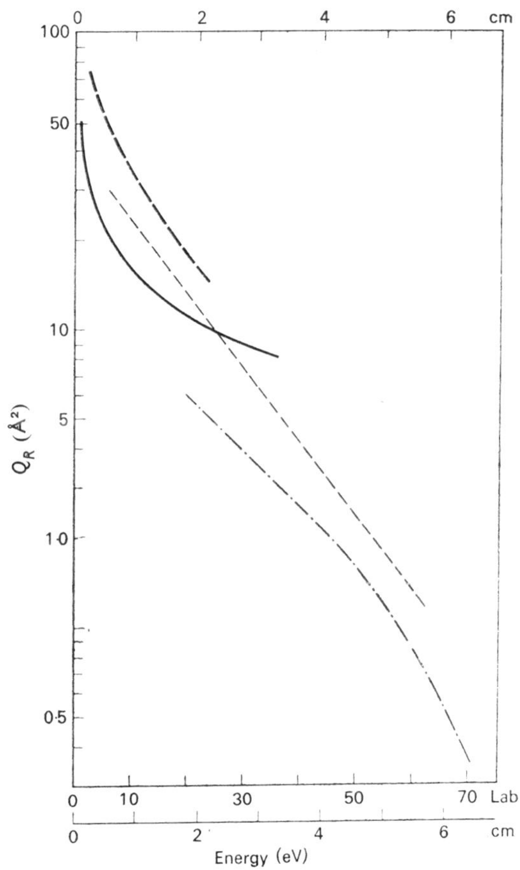

FIG. 9. Cross-section of $Ar^+ + D_2 \longrightarrow ArD^+ + D$ as a function of energy. ———— orbiting theory.[52] — — — — — Giese and Maier.[53] – – – – – – Homer, Lehrle, Robb, and Thomas.[3] —·—·—·—·— Lacmann and Henglein.[58]

produce a thin ribbon of secondary ions close to the backing plate (position 1) or close to the trapping grid (position 2). On optimizing all focus controls, the ions produced in position 1 arrive at the multiplier earlier than those produced in position 2 by up to one peak width, i.e. 40 nsec. The diffuseness of the primary beam at low energies will encompass both these extreme positions and thus reduce resolution by at least a factor of two.

Results obtained for the $N_2^+ + D_2 \longrightarrow N_2D^+ + D$ reaction by ourselves and three other groups of workers are compared in Fig. 8 with the cross-section variation predicted by eqn. (5). The orbiting theory can only be valid at energies sufficiently low for the impact parameter to be at least of the order of the sum of the van der Waals radii of the reactants. Other limitations of the theory have been discussed.[54] It appears that the theory underestimates the reaction cross-section at these lower energies. This is further in evidence in Fig. 9 which compares the data for the $Ar^+ + D_2 \longrightarrow ArD^+ + D$ reaction.

Henglein and Lacmann[57, 58] made a notable advance to our understanding of the mechanism of these processes in the higher energy range. They measured the velocity spectrum of the reactant and product ions and showed that, below a critical value at which reaction ceases, the reaction proceeds via a stripping mechanism where effectively one of the deuterium atoms is a "spectator" of the collision and does not gain momentum.

Neutral beams

Some studies have been made on the interaction of fast neutral beams with both gases and solids to yield ionic products. The neutral beams (of intensity equivalent to 10–1000 pA) are obtained through charge transfer.[59] The gas pressure between plates C and D (Fig. 2) is increased until ion gauge A reads approximately 5×10^{-5} torr. Highest yields are obtained through the use of symmetrical charge transfer reactions such as $Xe^+ + Xe \longrightarrow Xe + Xe^+$. The remaining ions in the beam are deflected out by a potential difference on the E_1–E_2 plates. The present apparatus is not well suited to neutral beam work since the increase of pressure in the primary source increases the impurity content in the reaction region R to about 10%, whereas in ion-impact studies the equivalent figure is $< 1\%$. The mechanism of production of the fast neutral beam is predominantly through charge transfer in the gas phase, and not electron tunnelling from the electrode surfaces, as is shown in Fig. 10 where the neutral beam intensity (as assessed by ionization of the detector gas methane) is proportional to the square of the pressure in the primary chamber.

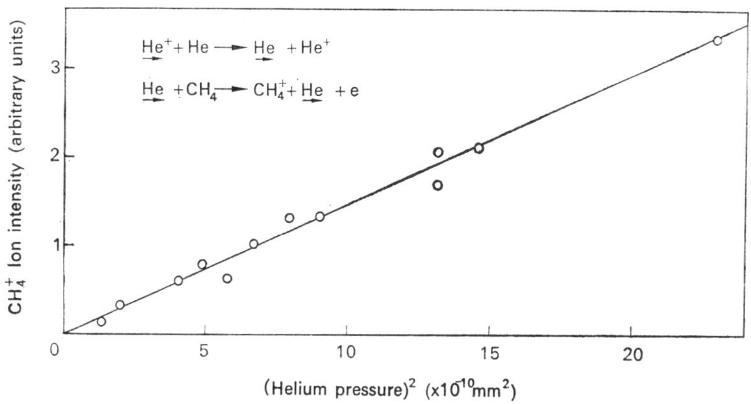

FIG. 10. Neutral beam intensity; dependence of methane ion intensity on helium pressure squared.

Major difficulties concerning neutral beam work are measurement of the neutral beam flux and assessment of the degree of translational energy inhomogeneity in the beam. Neutralization can occur at any point in the flight path and only the maximum translational energy may be specified. The fluxes are so low that the only feasible methods of measurement are through primary beam attenuation, slow ion production, or secondary electron emission.(60–62) The first two methods are based on the assumption that charge transfer, with little transfer of kinetic energy, is the predominant inelastic collision, and the last method assumes that the coefficient for kinetic ejection of electrons is the same for both ions and atoms of the same species and kinetic energy. Mahadevan *et al.*(62) however have shown that a discrepancy exists between the energy transfer properties of ions and neutral particles of the same species and kinetic energy due to the greater specular reflection of the neutrals compared with the ions.

The development of an electron bombardment detector for molecular beams has been described by Weiss.(63) This may lead to the removal of one present inadequacy, namely our complete lack of knowledge of the degree of excitation within the neutral beams, i.e. the type and percentage of long-lived metastable neutrals.

When polyatomic molecules are bombarded in the gas phase by neutral particles of energy up to 2 keV the relative abundances of the various ions produced differ from those obtained by bombardment of the same molecules by the corresponding ions of similar translational energy. This is illustrated in Fig. 11 which

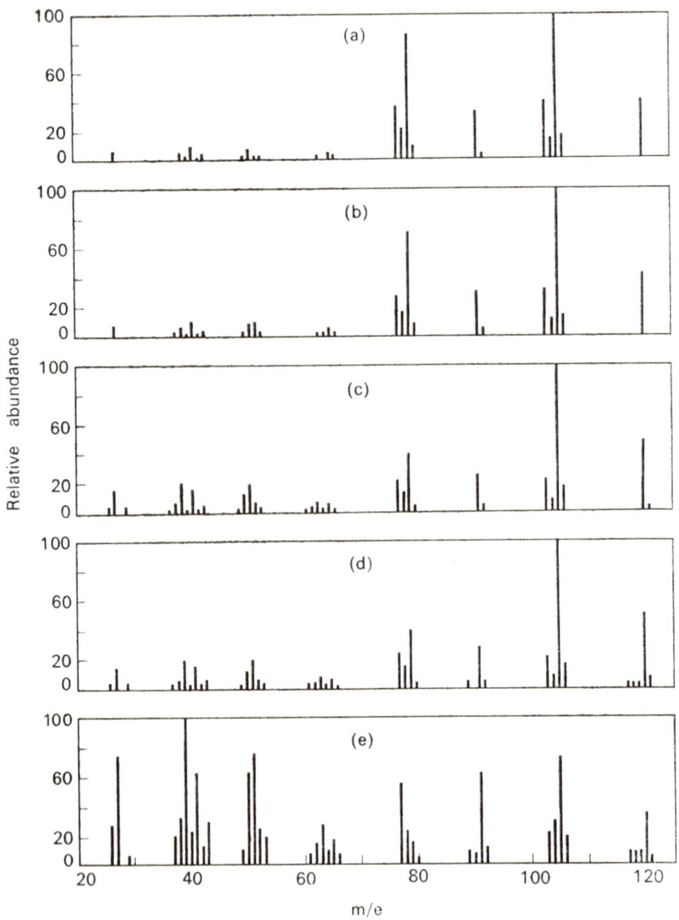

FIG. 11. Positive ion mass spectra of cumene on bombardment with nitrogen ions at (a) 50 eV, (b) 150 eV, (c) 1 keV, (d) 2 keV energy, and (e) nitrogen neutrals at 2 keV. (Only peaks $\geqslant 1\%$ of most intense ion peak are shown)

shows the cracking pattern of cumene resulting from bombardment by nitrogen ions and nitrogen neutrals. At low collision energies the energy transferred is predominantly the recombination energy of the nitrogen ion (15·3 eV) at high impact parameter with little transfer of momentum. There is consequently a certain selectivity of bond rupture. As the collision energy increases and the impact parameter decreases, the degree of fragmentation produced by ion impact approaches that produced by neutral impact. However, it can be seen that even at 2 keV energy, the contribution of the charge transfer process to the total ionization is still dominating.

Under similar conditions, neutral beams of argon, helium, and nitrogen readily ionize cumene (i.p. 8·76 eV) at translational energies of 250 eV but with methane (i.p. 13·1 eV) the signal for argon is several orders of magnitude lower compared with that for helium and nitrogen. This may be an indication that the neutral beams are in certain cases internally excited to long-lived metastable states. The time lapse between production and reaction is of the order of 0·5–5·0 μsec depending on the mass and energy. The potentials on the *CD* plates of Fig. 2 determine the maximum neutral beam translational energy. The influence of these potentials on the neutral impact mass spectrum of methane is shown in Table 1. It shows that either or both the translational energy of

TABLE 1. *Fragmentation of methane by neutral-particle bombardment*

Abundances obtained using the corresponding ion with the same translational energy are shown in parentheses

Neutral particle	Maximum translational energy (eV)	Product ion abundance				
		CH_4^+	CH_3^+	CH_2^+	CH^+	C^+
N_2	1500	18	100	34	12	–
He	2000	30 (2)	100 (19)	65 (100)	42 (43)	12 (10)
He	1500	25 (2)	100 (19)	67 (100)	45 (48)	12 (11)
He	1000	19 (2)	100 (16)	65 (100)	39 (48)	10 (12)
He	500	10 (2)	100 (16)	57 (100)	29 (52)	5 (13)
He	250	1 (1)	100 (14)	48 (100)	17 (57)	7 (15)

the neutrals and their flight times have a decided effect. CH_3^+ ions (a.p. 14·3 eV) are most abundant in both cases. Known metastable states for helium are at 19·81 eV (2^3S) and 20·61 (2^1S) and the highest known for N_2 is at 11·2 eV. Since argon has metastable states at 11·54 and 11·72 eV, a possible conclusion is that the fraction of the argon neutral beam in an excited state is considerably less than in the cases of helium and nitrogen. Also evident in the same table is that the variation of cracking pattern is much more marked in the case of neutral bombardment than for the corresponding ionic case. That it is purely a translational energy dependence cannot be inferred for the reasons stated above.

ION–SOLID INTERACTIONS

During the last decade, considerable progress has been made towards an understanding of the several processes involved in sputtering, particularly in the case of pure metals. Several reviews are available.[64–67] There are two basic aspects to the problem; one being concerned with the fate of the ions injected into a solid substrate and the consequent lattice alterations and bulk property changes. Such ion implantation (or ion doping) is of paramount interest in the field of semi-conductors. The initiation of chemical reactions in the solid state by accelerated ions, in contrast to γ-rays or neutrons, has received only slight attention.[68–69] The other aspect concerns the mechanism of removal or erosion of solids by both forward and back sputtering. A current useful application is in thin film preparation; whilst in the field of ion engine rocketry, efforts are being made to reduce the destructive effects of sputtering.

The complexity of the problem is illustrated in Fig. 12 where some of the parameters of interest in any ideal back-sputtering experiment are listed. One of the essential requirements is control of the background gas. The latter should ideally consist entirely of the primary gas with no impurity content. The choice of primary gas is dependent on whether or not strong adsorption and reactive sputtering is to be encouraged. The requirements of high intensity ion beams and low pressure in the target region demand a differential pumping system between ion source and target. Low pressure around the target is essential for any meaningful analysis of the sputtered particles since it assures a high mean free path of

BOMBARDING PARTICLES

Select: mass
charge
energy
angle of incidence
arrival rate

SPUTTERED PARTICLES

Determine: mass
charge
chemical formula
kinetic energy distribution
angular distribution
yield
threshold energies
degree of internal excitation

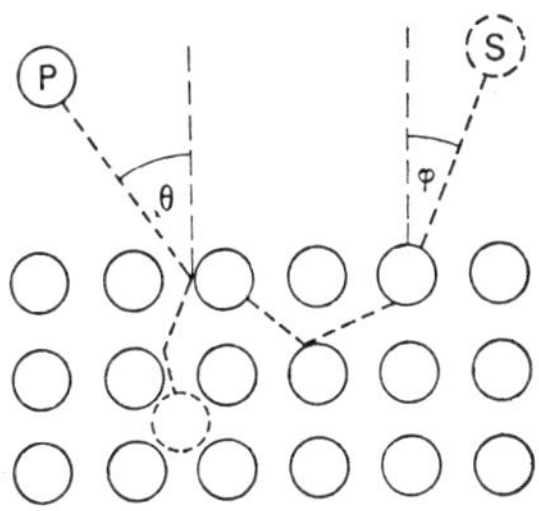

SURFACE AND BULK CONDITION

Temperature
Degree of order
single crystal face
polycrystalline
mixed structure (coal, polymer)
Optical examination
Bulk property changes
particle implantation

REQUIREMENTS

Background gas control — negligible pressure
— inert gas
— reactive gas
High intensity of bombarding beam compared with gas adsorption rate
Low pressure in vicinity of target—high mean free path of sputtered particles
Very high sensitivity of detection

FIG. 12. Schematic diagram of the experimental parameters in an ideal sputtering experiment.

the particles, with little probability of distortion of product ion abundances through collisions. The relationship between incident primary beam intensity and background gas adsorption, at a fixed temperature, has been assessed by Yonts and Harrison[70] and Honig.[71, 72]

Electric focusing is to be preferred to magnetic focusing, partly because greater proximity between ion source and target and hence greater intensity of primary ion bombardment may be achieved, and partly because mass discrimination in the analysis of secondary ions is reduced. The determination of threshold energies requires the use of ion beams of low energy (ca. 20 eV) and narrow energy spread. The retention of the same beam profile within a particular source geometry, as the beam energy is reduced, is most important. Variation of the profile can affect the observed threshold energy due to differing collection efficiences for secondary ions generated at different sites on the target.

Technique

One of several arrangements that we have employed in studies of sputtering is shown in Fig. 3 where the target filament is surrounded by a single box electrode which allows only ions sputtered in a direction within 10° of the normal to the surface to be identified and measured.

The mass resolution of the time-of-flight instrument is controlled predominantly by (a) the amplitude and rise time of the ion focus pulse, and (b) the degree of planarity and parallelism of the backing plate, trapping grid, focus grid, energy grid and multiplier dynode. The size of the extraction holes has only a weak effect, in direct contrast to magnetic analysers. The degree of resolution does depend to some extent on the initial velocity and direction of the ions before withdrawal into the flight tube. This is illustrated in Fig. 13. The primary Hg^+ ion beam at 20 eV translational energy is well collimated by the several focusing electrodes of the primary source and, before withdrawal into the flight tube, these ions have little lateral velocity parallel to the flight tube axis. Hg^+ ions produced by electron impact in the Bendix source (Fig 13b) have low thermal velocities in all directions, and some diffusion may occur in the short time before extraction occurs. The initial width of the ionization zone in this case is directly comparable with that of the primary beam at slit *D* since the defining slits were of the same dimensions. The Cu_2Cl^+ ions sputtered from copper chloride (Fig. 13c) are known, from retarding potential experiments, to have appreciable half-widths of kinetic energy spread ($\sim$2–4 eV) and, though the directional spread is restricted by a combination of size of repeller electrode slit and target to repeller slit distance, the effect of kinetic energy on the resolution is seen to be quite appreciable. This reduction of resolution is a severe limitation in studies of the sputtering of

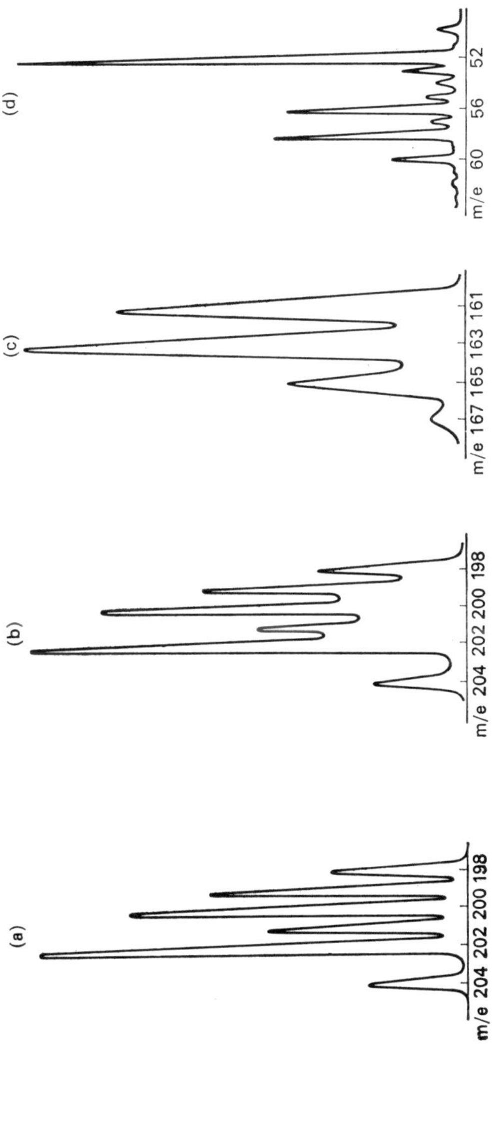

FIG. 13. Mass resolution of ions from four sources. (a) Hg^+ ions produced by electron impact in the auxiliary source. (b) Hg^+ ions produced by electron impact in the Bendix source. (c) Cu_2Cl^+ ions sputtered from cupric chloride. (d) Ni^+, Fe^+, Cr^+, and Mn^+ ions sputtered from a nichrome filament.

hydrogen-containing compounds but not of the equivalent fluoro-compounds or metals and inorganic salts.

The mass spectrum obtained from nichrome at 25°C with pulsed 2 keV N_2^+ ion beams (Fig. 13d) again illustrates the deterioration of resolution on sputtering. In this case gas adsorption is relatively low as is evident from the hydrocarbon peaks at *m/e* 55 and 57. In principle, the variation of both yield and mass spectrum of secondary ions from a target with temperature and rate of primary beam pulsing should give information on the accommodation coefficient of a gas. A necessary condition is that no interaction occurs between secondary particles, whether ionic or neutral, on ejection from a surface. There is evidence that this is not so in the case of hydrocarbon adsorption on nichrome.[73]

That the present system is satisfactory to the measurement of the kinetic energy distributions of the sputtered ions by a retarding potential technique has been assessed by Dillon *et al.*[73] in connection with some metal and inorganic salt studies. Further, variation of the target and repeller potentials with simultaneous current measurement at the target allows determination of the secondary electron yield coefficient, provided the reasonable assumption is made that the contribution of secondary positive and negative ions is negligible at bombarding energies below 2000 eV.

Metals

At a given pressure, the degree of surface contamination due to adsorption may be decreased by increasing either the bombarding ion intensity or the temperature of the target sample. The effect of ion intensity is illustrated in Fig. 14 in a typical case of metal sputtering under poor vacuum conditions. The base metal aluminium contains sodium and potassium as impurities and, in the continuous mode of operation, in addition to the major ions Al^+, Na^+ and K^+, the minor peaks due to H_2O^+, $AlOH^+$, $C_2H_5^+$ and $C_3H_7^+$ are recorded. The contribution to the spectrum of ions arising from dissociative charge transfer between the primary ions and the background gases is readily shown to be negligible by deflecting the primary beam slightly off target. No complex ions such as NAl^+ or N_2Al^+ characteristic of both the bombarding gas and base metal are detected. On switching to very short duration pulse widths ($\sim$50 nsec) at 10 kc/s repetition frequency the ion peaks arising from hydrocarbon adsorption become dominant over the base metal peaks.

In general the impurity species most readily detected as positive ions are those whose ionization potentials are lowest, e.g. the alkali and alkaline earth metals.

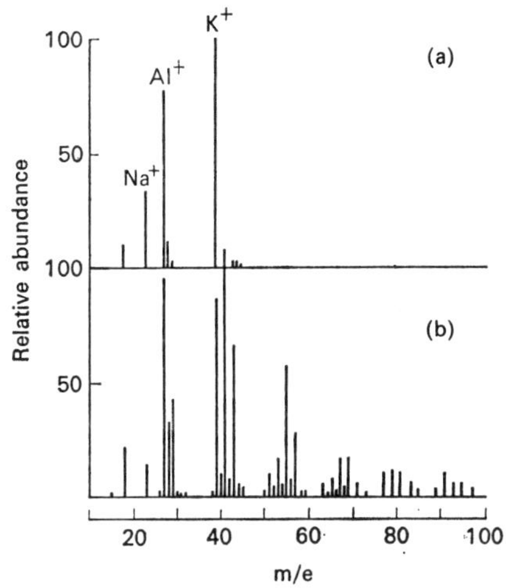

FIG. 14. Effect of bombarding current intensity on the mass spectrum of positive ions sputtered from aluminium at 20°C by 2 keV N_2^+ ions. Arrival rate of primary ions in the continuous mode (a) is 100× that in the pulsed mode (b).

Prominent on occasion are the complex ions of the base metal with water and the hydroxyl radical to give ions such as $[M(H_2O)_nOH]^+$ where M is a metal atom of any valency and $n \leqslant 3$.

The effect of heat alone is illustrated in Fig. 15. Mass spectra of the sputtered ions were obtained as the temperature of the nichrome filament was raised to 400°C and then returned to 20°C. The hydrocarbon ions disappear and then reappear in much the same relative yields. Noticeable, however, is an increase in the surface concentration of manganese and iron. The relative amounts of the single metal ions Ni^+, Fe^+, Cr^+ and Mn^+ sputtered is a strong function of temperature. The same is true of dimer ions such as Ni_2^+ and $NiFe^+$. Some of the factors which influence the mass spectra of alloys sputtered are mass and nature of the bombarding ion, kinetic energy of the bombarding ion, ionization potential and size of the atom ejected, and

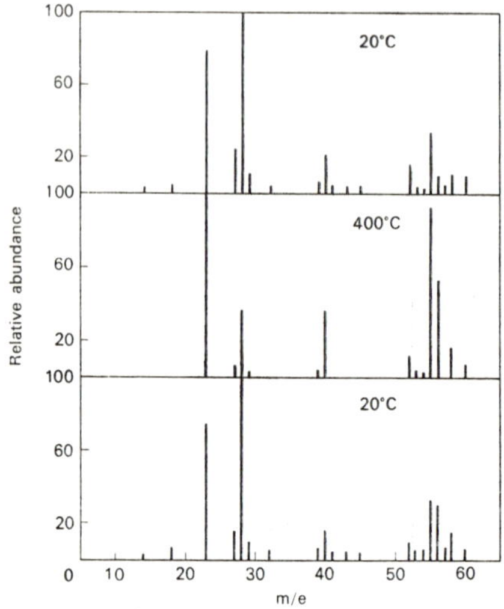

FIG. 15. Effect of temperature on the mass spectrum of positive ions sputtered from nichrome. Bulk analysis: 65% Ni, 15% Cr, 25% Fe.

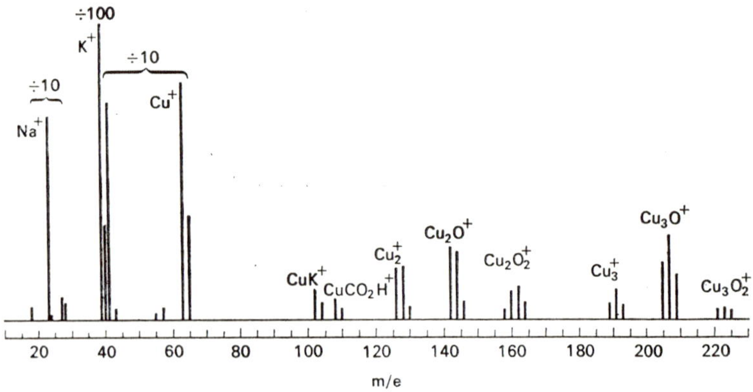

FIG. 16. Complex ions sputtered from an oxidized copper surface. Bombarding beam: microsec pulses of Ar^+ ions at 2 keV energy. Temperature 150°C.

background gas pressure. The analytical possibilities of the method for bulk analysis have been explored by Dillon *et al.*[73]

In addition to the detection of chemisorbed species and bulk analysis, the ion bombardment technique may prove useful in the analysis of surface corrosion. The change in mass spectrum as the surface layers are gradually sputtered away may be followed. Figure 16 shows, for instance, the condition of a copper surface after oxidation. There is copious yield of various copper oxide ions and, compared with the spectrum of pure copper, the yield of the metal ions Cu_2^+ and Cu_3^+ has increased by orders of magnitude. The peaks at m/e 108 and 110 are probably due to $^{63}CuCO_2H^+$ and $^{65}CuCO_2H^+$. The hydrocarbon and water peaks at m/e 55, 53, 43, 27 and 18 arise to some extent from dissociative charge transfer in the gas phase.

Polymers

Mass analyses of the positive ions ejected from various polymers on bombardment with rare gas and nitrogen ions have been made.[74] The deleterious effect of surface charge-up was largely overcome through the use of very thin polymer films and short duration ion pulses. An alternative approach which has been used involves bombardment with the corresponding neutral beams.[1, 74]

Figure 17 shows some typical positive ion mass spectra obtained in the initial stages of damage, i.e. when the total dose is in the range 10^{10}–10^{13} ions cm^{-2}. (Damage may be detected optically after doses in the range 10^{13}–10^{15} ions cm^{-2}.) High mass fragment ions, characteristic of the base polymer, are sputtered off. Thus in the case of polyethyleneterephthallate the ion of m/e 252 is probably O—M—CH_2—CH_2—O^+ and m/e 281 is O—M—CH_2——CH_2—O—CO—H^+, where M is the basic polymer unit m/e 192, [—CH_2—CH_2—O—CO—Ph—CO—O—]. The major ion sputtered is m/e 73, CH_2—CH_2—O—COH^+. Nylon-6 gives a high yield of a trimer unit at m/e 342.

Polytetrafluoroethylene was studied in more detail and the influence of primary ion type, incident ion current density, primary beam energy and temperature on the type, yield, and kinetic energy of ion ejected was determined. The retarding potential curves for the three ions CF^+, CF_3^+ and $C_3F_5^+$ were found to be similar, with an energy value of 2–3 eV at half height for bom-

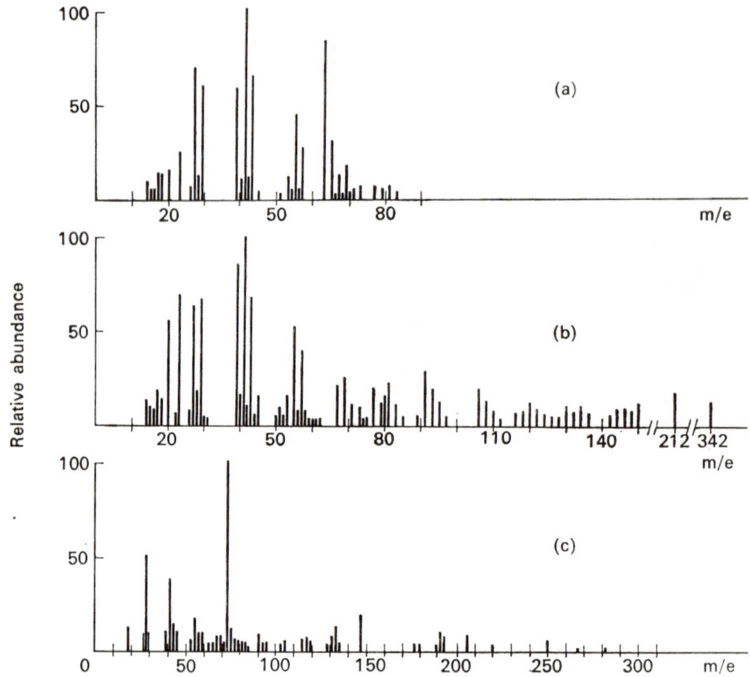

FIG. 17. Positive ion mass spectra of three polymers sputtered at room temperature. (a) Polypropylene; 2 keV Ne^+ continuous beam. (b) Nylon-6; 2 keV Ne^+ continuous beam. (c) Polyethylene terephthalate; 2 keV N_2^+ pulsed beam. (Only peaks $\geqslant$ 2% of most intense ion peak are shown.)

barding energies of 1–2 keV. The mass spectra are weakly dependent on primary ion mass and recombination energy, but the influence of primary ion on yields decreased in the order $Ne^+ > N_2^+ > Ar^+ > Xe^+$. With the apparatus which has been described it has been possible to study the purely thermal degradation of the same polymers using either electron or ion impact to detect the neutral species desorbed.

REFERENCES

1. LEHRLE, R. S., ROBB, J. C., and THOMAS, D. W., *J. Sci. Instr.* **39,** 458 (1962).
2. HOMER, J. B., LEHRLE, R. S., ROBB, J. C., TAKAHASI, M., and THOMAS, D. W., *Adv. in Mass Spectrometry* **2,** 503 (1962).

3. HOMER, J. B., LEHRLE, R. S., ROBB, J. C., and THOMAS, D. W., *Adv. in Mass Spectrometry* **3,** 415 (1964).
4. LEHRLE, R. S., ROBB, J. C., SCARBOROUGH, J., and THOMAS, D. W., paper in *Adv. in Mass Spectrometry*, Vol. 4, 687 (1968).
5. LINDHOLM, E., *Proc. Phys. Soc.* **A66,** 1068 (1953).
6. LINDHOLM, E., *Z. Naturf.* **9a,** 535 (1954).
7. TALROZE, V. L., *Izvest. Akad. Nauk. SSR*, ser. fiz. **24,** 1001 (1960).
8. TALROZE, V. L., *Pure App. Chem.* **5,** 455 (1962).
9. (a) GUSTAFSSON, E., and LINDHOLM, E., *Ark. Fysik* **18,** 219 (1960). (b) VON KOCH, H., and LINDHOLM, E., *Ark. Fysik* **19,** 123 (1961).
10. LAMPE, F. W., FIELD, F. H., FRANKLIN, J. L., *Adv. in Reaction Kinetics* **1,** 67 (1961).
11. TALROZE, V. L., and FRANKEVICH, E. L., *Zh. Fiz. Khim.* **34,** 2709 (1960).
12. WAHRHAFTIG, A. L., *Adv. in Mass Spectrometry* **1,** 274 (1959).
13. SHANNON, T. W., and HARRISON, A. G., *J. Chem. Phys.* **43,** 4201 (1965).
14. SHANNON, T. W., MEYER, F., and HARRISON, A. G. *Can. J. Chem.* **43,** 159 (1965).
15. HAND, C. W., and VON WEYSSENHOFF, H., *Can. J. Chem.* **42,** 195 (1964).
16. FUTTRELL, J. H., and RYAN, K. R., *J. Chem. Phys.* **42,** 824 (1965).
17. FUTTRELL, J. H., and RYAN, K. R., *J. Chem. Phys.* **43,** 3009 (1965).
18. CERMAK, V., and HERMAN, Z., *Nucleonics* **19,** 106 (1961).
19. TIERNAN, T. O., and FUTTRELL, J. H., *J. Chem. Phys.* **39,** 2539 (1963).
20. TRUJILLO, S. M., ROTHE, E. W., and NEYNABER, R. H., *Rev. Sci. Instr.* **37**, 1655 (1966).
21. ISHII, H., and NAKAYAMA, K., *Trans. 8th AVS Nat. Vac. Symposium* **1,** 519, 1961 (Pergamon Press, 1962).
22. ROTHE, E. W., *J. Vac. Sci. Tech.* **1,** 66 (1964).
23. DE VRIES, A. E., and ROL, P. K., *Vacuum* **15,** 135 (1965).
24. HARRISON, A. G., JONES, E. G., GUPTA, S. K., and NAGY, G. P., *Can. J. Chem.* **44,** 1967 (1966).
25. OTVOS, J. W., and STEVENSON, D. P., *JACS* **78,** 546 (1956).
26. LAMPE, F. W., FIELD, F. H., and FRANKLIN, J. L., *JACS* **79,** 6129 (1957).
27. WILEY, W. C. and MCLAREN, I. H. *Rev. Sci. Instr.* **26,** 1150 (1955).
28. HASTED, J. B., *Proc. Roy. Soc.* A**205,** 421 (1951).
29. MASSEY, H. S. W., and BURHOP, E. H. S., *Electronic and Ionic Impact Phenomena* (Oxford, Univ. Press, 1952), p. 265.
30. LINDHOLM, E., *Rev. Sci. Instr.* **31,** 210 (1960).
31. FUTTRELL, J. H., *Adv. in Mass Spectrometry* **3,** 432 (1965).
32 FUTTRELL, J. H., and TIERNAN, T. O., *J. Chem. Phys.* **39,** 2539 (1963).
33. INGHRAM, M. G., and HAYDEN, R. J. *A Handbook on Mass Spectrometry*, Nuclear Sci. Ser. No. 14 1954 (Nat. Acad. Sci.—Nat. Res. Council Publn. 311), pp. 41–44.
34. HOFFERT, F., PAULUS, J., and ADLOFF, J. P., *Rev. phys. appliquée* **1,** 43 (1966).
35. LA LAU, C., private communication.
36. HOMER, J. B., LEHRLE, R. S., ROBB, J. C., and THOMAS, D. W., *Trans. Faraday Soc.* **62,** 619 (1966).
37. MASSEY, H. S. W., and BURHOP, E. H. S., *Electronic and Ionic Impact Phenomena* (Oxford Univ. Press, 1952), p. 441.
38. HASTED, J. B., *Adv. Electronics and Electron Physics* **13,** 1 (1960).

39. Sena, L. A., *Zhur. eksp. teor. fiz.* **9,** 1320 (1939).
40. Demkov, U. H., *Ann. Lenigr. Univ.* **146,** 74 (1952).
41. Hagstrum, H. D., *Phys. Rev.* **104,** 309 (1956).
42. McGowan, J. W., and Kerwin, L., *Can. J. Phys.* **41,** 1535 (1963).
43. Fogel, I. M., Kozlov, V. F., Kalymkov, A. A., and Muratov, I. M., *J. Exp. Theor. Phys.* **36,** 929 (1959).
44. Utterback, N. G., and Amme, R. C., *Atomic Collision Processes*, Ed. M. R. C. McDowell (North Holland, 1964), pp. 847–53.
45. McGowan, J. W., and Kerwin, L., *Can. J. Phys.* **42,** 2086 (1964).
46. McGowan, J. W., Marmet, P., and Kerwin, L., *Atomic Collision Processes* Ed. M. R. C. McDowell (North Holland, 1964), pp. 854–61.
47. Giese, C. F., and Maier, W. B. (II), *J. Chem. Phys.* **39,** 197 (1963).
48. Maier, W. B. (II), *J. Chem. Phys.* **41,** 2174 (1964).
49. Maier, W. B. (II), *J. Chem. Phys.* **42,** 1790 (1965).
50. Light, J. C., *J. Chem. Phys.* **40,** 3221 (1964).
51. Scarborough, J., Ph.D. thesis, University of Birmingham, 1966.
52. Gioumousis, G., and Stevenson, D. P., *J. Chem. Phys.* **29,** 294 (1958).
53. Giese, C. F., and Maier, W. B., *J. Chem. Phys.* **39,** 739 (1963).
54. Homer, J. B., Lehrle, R. S., Robb, J. C., and Thomas, D. W., *Nature* **202,** 795 (1964).
55. Henglein, A., Lacmann, K., and Jacobs, G., *Ber. Bursenges* **69,** 279 (1965).
56. Turner, B. R., Fineman, M. A., and Stebbings, R. F., *J. Chem. Phys.* **42,** 4088 (1965).
57. Henglein, A., and Lacmann, K., *Ber. Bunsenges Physik. Chem.* **69,** 292 (1965).
58. Henglein, A., and Lacmann, K., *Ber. Bunsenges* **69,** 286 (1965).
59. Amdur, I., Glick, C. F., and Pearlman, H., *Proc. Am. Acad. Arts Sci.* **76,** 101 (1948).
60. Utterback, N. G., and Miller, G. H., *Rev. Sci. Instr.* **32,** 1101 (1961).
61. Utterback, N. G., and Miller, G. H., *Phys. Rev.* **124,** 1477 (1961).
62. Mahadevan, P., Layton, J. K., Comeaux, A. R., and Medved, D. B., *J. App. Phys.* **34,** 2810 (1963).
63. Weiss, R., *Rev. Sci. Instr.* **32,** 397 (1961).
64. Wehner, G. K., *Adv. in Electronics and Electron Phys.* **7,** 239 (1955).
65. Colligon, J. J., *Vacuum* **11,** 272 (1961).
66. Kaminsky, M., *Atomic and Ionic Impact Phenomena on metal Surfaces* (Berlin, Springer-Verlag, 1965).
67. Chaddeston, L. T., *Radiation Damage in Crystals*, (Methuen, 1965).
68. Logan, S. R., and Moore, W. J., *J. Phys. Chem.* **67,** 1042 (1963).
69. Anderson, T., and Sørensen, G., *Trans. Faraday Soc.* **62,** 3427 (1966).
70. Yonts, O. C., and Harrison, D. E., *J. Appl. Phys.* **31,** 1583 (1960).
71. Honig, R. E., *Adv. in Mass Spectrometry* **1,** 162 (1959).
72. Honig, R. E., *Adv. in Mass Spectrometry* **2,** 25 (1962).
73. Dillon, A. F., Robb, J. C., and Thomas, D. W., to be published.
74. Dillon, A. F., Lehrle, R. S., Robb, J. C., and Thomas, D. W., *Advances in Mass Spectrometry*, Vol. 4, 477 (1968).

HIGH TEMPERATURE STUDIES OF INORGANIC SOLIDS

R. Bowles
Central Electricity Research Laboratories, Leatherhead, England

INTRODUCTION

In the determination of vapour pressures, heats of vaporization, dissociation and atomization of inorganic substances, a variety of classical techniques may be used. Manometric vapour pressure and decomposition pressure measurements, transport and Knudsen vaporization measurements are examples of techniques which while providing very useful information, often suffer the disadvantage that only a "gross" chemical effect is being measured. The need to isolate and identify the reactions which can occur within a system at a high temperature has been met by the use of mass spectrometry.

Any review of high temperature mass spectrometry cannot be confined entirely to results from time-of-flight (TOF) machines. The reader must be referred to earlier comprehensive reviews on the subject.[1–3] He will then appreciate the considerable advance in the understanding of high temperature reactions which has been made in the last 15 years. It is only in the past 7 years that owners of TOF mass spectrometers have been making their contributions in this field with the study of the vaporization of solids under Langmuir and Knudsen conditions. Very recently the fast response capability of TOF machines has been utilized to study vaporization using lasers and shock-heated filaments.

KNUDSEN CELL INLET SYSTEMS

Inside a Knudsen cell (KC) an equilibrium is set up between the vapour and the condensed phase of the sample. The vapour effuses from a small orifice in the top of the cell and passes into the mass spectrometer source where it is ionized. Essential characteristics of KC's are discussed by Nesmeyanov.[8]

The Bendix TOF mass spectrometer with its readily accessible ion source can, with relative ease, be fitted with a high temperature inlet system. Four systems, which are all basically similar,

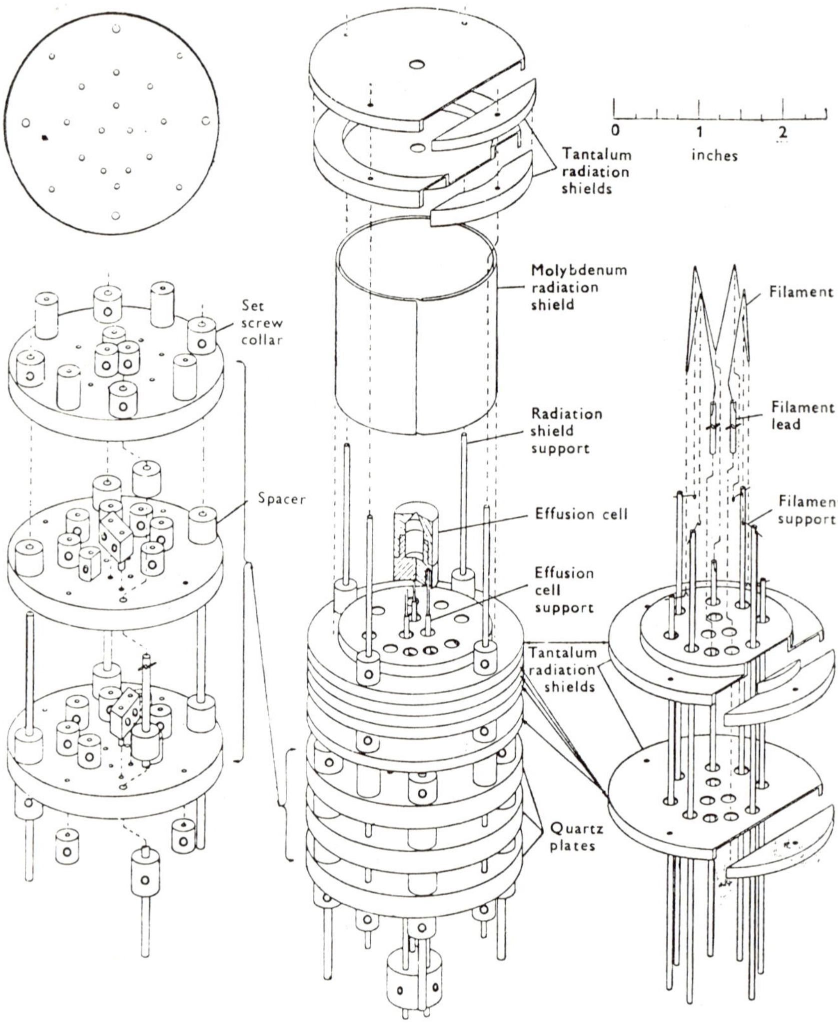

FIG. 1. Knudsen cell furnace assembly developed by Rauh, Sadler and Thorn.[5] (By permission of the Argonne Laboratory, Argonne, Illinois.)

FIG. 2. Tower assembly of Knudsen cell inlet system. (By permission of the Bendix Corp.)

have been described in the literature.[4–7] Each uses a vertical molecular beam (X-axis) with the electron and ion beams mutually perpendicular (Y- and Z-axes). This particular orientation is now generally used. An example of a furnace design developed at the Argonne National Laboratories, USA, is shown in Fig. 1. The Bendix designed inlet system has incorporated the important features of earlier designs and therefore it will be described in more detail.

The system is shown in Figs. 2 and 3. The cell or crucible is supported on a tripod of three pointed tungsten rods. One of the rods may be replaced by a tungsten 5% rhenium, tungsten 26% rhenium thermocouple encased in a tantalum and ceramic sheath. Two sets of concentric heat shields surround the crucible. The shield top caps have central slits to pass the molecular beam from the cell. The beam is aligned by moving the tripod assembly with an external screw adjustment to one of the legs. The crucible is normally heated by two tungsten filaments (0·01 in. diameter) suspended between tantalum electrodes. These filaments will take a maximum of 7·5 A and provide radiation heating up to ca. 1170°K. Higher temperatures up to 2600°K are achieved with a combination of radiation and electron bombardment heating. In bombardment heating a 1100 V d.c. potential is set up between the filaments and the crucible at ground potential. It is therefore necessary to electrically insulate the heat shields from ground using, in this case, boron nitride supports.

The assembly is surrounded by a water-cooled stainless steel jacket, and the chamber is pumped through a 2 in. port. For speed of operation and reduction of background spectra it is essential to use a fast pumping system. A 2 in. oil diffusion pump (Edwards E02) with water-cooled baffle and vapour trap mounted directly on to the side of the chamber gives a satisfactory pumping speed of 15 l./sec at the cell. This allows sample changing and high temperature degassing to be carried out in a normal working day. It is also important to have reasonably fine control on the vacuum valves so that powdered samples may be degassed slowly.

The electron bombardment heating control system can only operate satisfactorily at chamber pressures below 10^{-5} torr. Figure 4 shows the effect of chamber pressure on the crucible temperature at a constant heating power control setting. At pressures above 10^{-5} torr a positive feed back situation develops in the normally stabilized control system. As pressures are higher in the immediate area of the cell the loss of control is probably due to the onset of a glow discharge between the filaments and the crucible. The chamber pressure must therefore be maintained in

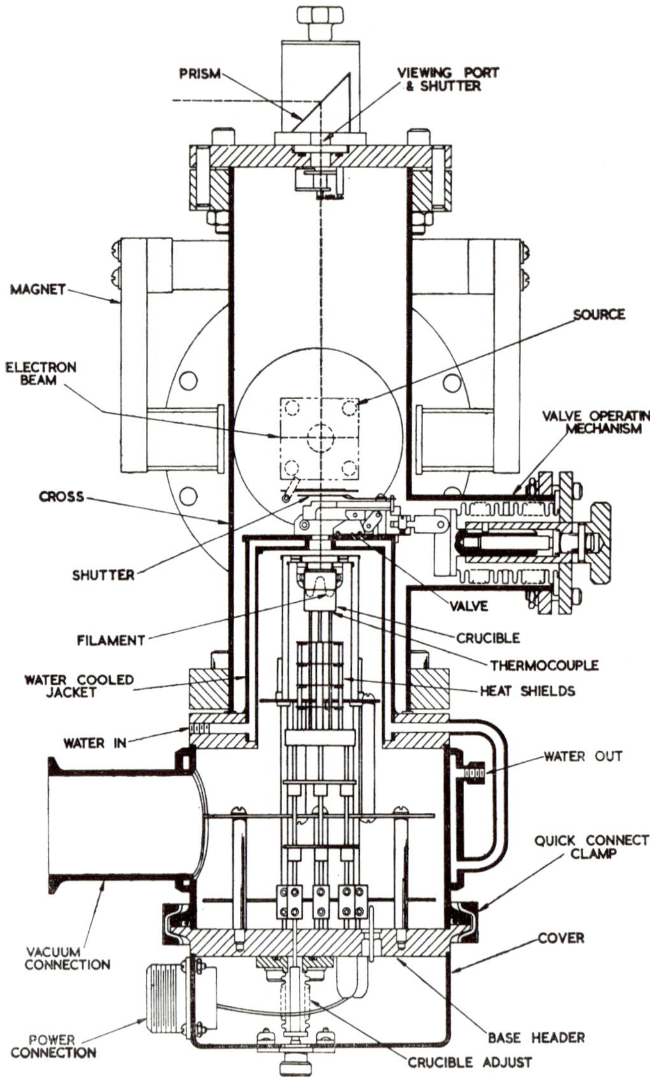

FIG. 3. Bendix designed Knudsen cell furnace assembly.

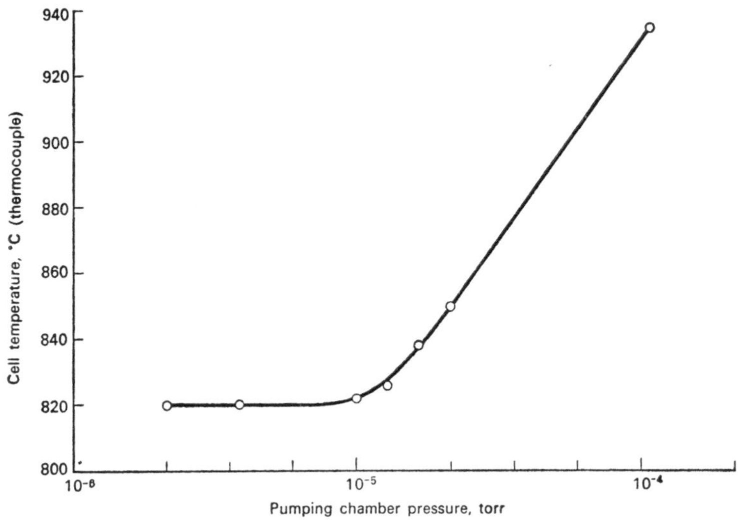

FIG. 4. Variation of electron bombardment heating effect with pressure.

the 10^{-} –10^{-7} region. The heating system is protected from sudden pressure rises by an overload cut-out on the vacuum gauge circuit.

The bombardment electrons can, under certain circumstances, ionize the sample molecules. This effect was particularly noticeable in handling potassium compounds, some of which decompose to give readily ionized potassium vapour. To minimize the effect the normal maximum temperature of radiation heating can easily be extended by inserting a further pair of heating filaments between electrodes 1, 4 and 2, 3, in parallel with the standard pair. These filaments are conveniently fixed through the holes in the top of the support rods and clamped with small tantalum pegs. The standard heating system is then bypassed with a controlled source of a.c. heating current of up to 15 A. This will give a maximum temperature of about 1400°K and usefully extends the range of samples which may be heated purely by radiation. The four filament arrangement completely surrounds the crucible and gives a more uniform heat distribution. The system must be returned to the standard configuration when electron bombardment heating is used. An alternative system to reduce thermal gradients is to heat the top and bottom of the cell separately using two independently controlled filaments.(7)

THE DETERMINATION OF VAPOUR PRESSURE

The pressure P_i of a species i effusing from a Knudsen cell is given by the equation[(3)]

$$P_i = \left(\frac{3\pi^2 L^2 k}{2i^- l\eta s}\right)\frac{I_i^+ T}{\sigma_i \gamma_i}, \tag{1}$$

where L is the distance between the orifice and the centre of the ionizing region normal to the orifice, k the constant from the ideal gas equation, i^- the intensity of the electron beam, l, the length of the path at crossing of electron and molecular beam, η the yield or transmission of the mass spectrometer, s the area of the KC orifice, I_i^+ the positive ion current from species i, T the absolute temperature, σ_i the ionization cross-section for species i, γ_i the gain of the electron multiplier for ion i.

The factor inside the brackets of eq. (1) is termed the sensitivity S of the mass spectrometer and is constant for a given set of experimental conditions. The sensitivity cannot easily be calculated and is normally determined experimentally.

The pressure P_i may be determined directly provided the substance does not decompose or evaporate to give a high proportion of minor species. A known mass is evaporated in a time t with the mass spectrometer serving purely to check the total evaporating time. The vapour pressure is then calculated from the ideal Knudsen effusion equation (2):

$$P = \frac{Z(MT)^{1/2}}{44{\cdot}33\,Kat,} \tag{2}$$

where P is the vapour pressure in atmospheres, Z the total moles of material lost in the time t secs, M the molecular weight of the vapour, a the orifice area in cm^2, K, the Clausing factor, which is the ratio of the effective pumping speed through the orifice to the ideal pumping speed through an orifice of equal size in an infinitely thin plate. Methods exist for calculating K for most common shapes of orifice.[(10, 11)]

To determine P_i for each species in a more complex system the factor S must be determined. One method used frequently is to sample a well-defined homogeneous or heterogeneous equilibrium. It is convenient to measure the ion intensity produced by a sample of silver or gold at the melting point and relate this to the known vapour pressure.[1, 3]

The ion current I_i^+ and temperature T are determined directly which leaves the factors σ_i and γ_i to be measured or estimated.

Experimental values of ionization cross-sections are now available for most of the common elements.[9] In the past the majority of workers have required only relative ionization cross-sections and the calculated values of Otvos and Stevenson[12] have been used almost universally. However, when calculated on an absolute basis by the choice of a suitable standard, Otvos and Stevenson values were only in fair agreement with experimental data.[13] It was obvious that some refinement of these theoretical calculations was required. Recently Mann[14] has revised the cross-sectional data for atoms using a method of calculation similar to those of Otvos and Stevenson, based on the summation of the mean square orbital radii of outer electrons. Stafford[15] has used a different method based on ionization potentials and has calculated cross-sections for most of the common elements. Taken as a whole the recent values are in better agreement with experiment, although discrepancies still exist. For the moment it would seem advisable to use reliable experimental values, if available, or one only of the sets of theoretical values until the whole field is critically reviewed.

For polyatomic molecules Otvos and Stevenson proposed the "additivity rule" which states that the molecular ionization cross-section is the sum of the atomic cross-sections. However, in a number of investigations it has been shown that agreement between experimental and predicted cross-sections using the additive principle is far from good,[9] although satisfactory correlation has been obtained for the alkanes and some simple molecules such as nitrogen and oxygen. When cross-sectional data is available for a similar group of molecules it is possible to obtain an estimate by an empirical correlation. When this data is not available there is as yet no practical alternative to using the additive principle although at best this can only be regarded as a semi-quantitative method of estimation.

It is often desirable to ionize the molecular beam with electrons of low energy either to prevent fragmentation of the molecular ions or to increase the molecular beam/background ratio, since in general the background consists of hydrocarbon ions with appearance potentials of 14 eV or more, whereas the inorganic species in the molecular beam usually have ionization potentials in the 10 eV region. Now the theoretical ionization cross-section for single ionization, σ_{max} is calculated for 60–75 eV electrons. If in practice low energy (10–20 eV) electrons are used σ_{max} must be corrected to give σ_E, the cross-section at an electron energy of E volts. One factor frequently used is given in eqn.[16] (3)

$$\sigma_E = \sigma_{max}\left(\frac{E-A}{Vm-A}\right), \tag{3}$$

where A is the appearance potential of the ion, Vm the electron energy at which the cross-section reaches a maximum.

Equation (3) is only applicable in the low energy region where the ion efficiency curve is approximately linear. In the higher energy region dissociative ionization may occur which will effectively reduce the actual cross-section for the production of the molecular ion below that predicted by eqn. (3).

The multiplier efficiency γ_i [eqn. (1)] is to a near approximation inversely proportional to the square root of the mass M of the ion. Kant has shown that this relationship holds for the inert gases argon, krypton and xenon in the TOFMS.[17] For polyatomic ions there are also molecular effects on the multiplier gain due, it is thought, to the ions breaking up at the surface of the multiplier.[18] An additional correction factor for this effect is shown in eqn.[3] (4):

$$\gamma_i \approx \frac{1}{M^{1/2}}\left[1+\frac{\dfrac{\Sigma m_i^{1/2}}{(\Sigma m_i)^{1/2}}-1}{3}\right] \tag{4}$$

where Σm_i = sum of the atomic masses.

THERMODYNAMIC CONSIDERATIONS

The change of free energy ΔG°_T relating to a physical or chemical change or reaction is given by

$$\Delta G^\circ_T = -RT \ln Kp = \Delta H^\circ_T - T\Delta S^\circ_T, \tag{5}$$

where Kp is the equilibrium constant of the reaction, ΔH°_T the standard enthalpy change at a temperature T, and ΔS°_T the corresponding entropy change.

Consider the determination of the heat of vaporization of:

$$\text{A (solid, liquid)} \rightleftharpoons \text{A (gas)}.$$

Now eqn. (1) simplifies to $P_i = k'IT$, where k' is a constant (provided the factors γ_i and σ_i do not vary with temperature). For the vaporization process it follows that

$$\ln Kp = \ln P_A = \ln (I_A T) + \text{constant}, \qquad (6)$$

where I_A is the ion current due to A (gas).

Combining eqn. (5) and (6) we have

$$\ln Kp = \ln I_A T + \text{constant} = -\frac{\Delta H^\circ_T}{RT} + \frac{\Delta S^\circ_T}{R}. \qquad (7)$$

Thus a plot of $\ln (I_A T)$ against $1/T$ will give a line of slope $-\Delta H^\circ_T/R$.

A second method of determining ΔH°_T requires the absolute determination of Kp at a single temperature T and a knowledge of ΔS°_T. Values of standard entropy are known for the elements and for many simple inorganic compounds or may be calculated from heat capacity data.

SOME GENERAL CONSIDERATIONS OF THE MEASURING TECHNIQUE

Standardization experiments

In a pressure determination the standard metal should, if possible, be vaporized in the same experiment as the substance under investigation. If the standard and test samples react when heated together then a divided cell may be used. Each half of this cell has a separate orifice and the body of it is tilted to select one or other of the molecular beams.

If standardization is performed as a separate experiment, then the unknown pressure must be determined under identical instrumental conditions. It is not however easy to ensure constant sensitivity, where each experiment may take a working day to complete. It is better in this case to relate the primary calibration to a second experiment using an inert gas, e.g. argon. A gas is less difficult to handle and a secondary calibration of this kind may be performed immediately before the main pressure determination.

With the TOFMS the gas may be introduced into the K.C. pumping chamber at room temperature, through the air-bleed valve, and is allowed to flow through the molecular beam slit into the spectrometer source. Most of the gas is pumped away by the KC pumping system and the equilibrium pressure within the chamber is measured with the calibrated Bayard–Alpert ionization gauge. The A^{40} ion peak is monitored at say 75 eV, where the variation of σ with electron energy is small. Apart from electron energy which may be precisely altered and measured (with a digital voltmeter) all other instrumental parameters, e.g. trap current, multiplier gain are maintained constant for the main pressure determination which follows, and which is related, through the purely relative argon determination, to the primary standardization.

The sensitivity of the TOFMS depends on the control settings for the ion source, flight tube, electron multiplier and analogue electrometer. Particular attention should be paid to the width of the electrometer "gate" pulse. When constantly monitoring an ion peak a wide gate pulse should be used to cover the whole of the ion peak. *A* large and erroneous change in ion current will be produced if the pulse moves in time relative to the ion peak. This is prevented by ensuring the electronic stability of the gate pulse timing circuit.

Temperature measurement

The tungsten–rhenium thermocouple usefully covers the whole operational temperature range, although if used above 2000°C it may require recalibration. For the lower temperature range, using purely radiation heating a platinum, platinum–rhodium thermocouple is preferred especially where it can be directly welded to a platinum crucible. The output of the thermocouple may be fed into a suitable high impedance digital voltmeter which then provides continuous temperature readings in mV.

Cell temperatures in the range 800–2200°C can be measured

using a disappearing filament pyrometer. The orifice of the cell may be observed through the viewing port and prism shown in Fig. 3. The difficulty of viewing an orifice ca. 0·5 mm diameter at a minimum distance of 25 cm is overcome by using a micro-pyrometer (e.g. Pyro-werk GMBH micro-pyrometer). A word of caution should be expressed here. These instruments have a high

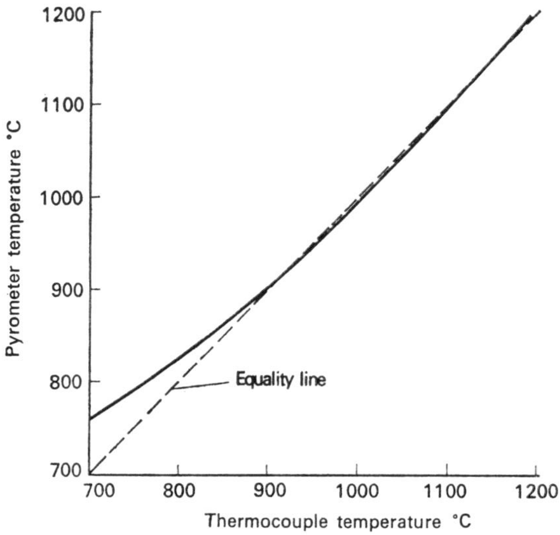

FIG. 5. Comparison between pyrometer and thermocouple.

magnification and hence a small field of view. It is necessary therefore to ensure very careful optical alignment, otherwise erroneously low temperature readings are obtained. It is also essential that the full cone of radiation emitted from the cell orifice reaches the pyrometer objective lens. If an orifice $>0{\cdot}5$ mm diameter is used then it is necessary to enlarge the slits in the water cooled jacket and the molecular beam collimating plate on the ionizing source.

The pyrometer and platinum–rhodium thermocouple have a useful overlap of temperature range and a comparison may be made between them, as is shown in Fig. 5. At the low temperature end there is a positive deviation of pyrometer temperature reading which has also been observed by other workers.[(4)] The error almost certainly lies in the pyrometer readings and may be caused by reflected radiation from the hot filaments.

SOME PROBLEMS OF STUDYING DECOMPOSITION REACTIONS

Increasing attention is being paid to compounds which evaporate and then decompose in the gas phase. The decomposition reaction may be slow compared with the rate of evaporation of the complete molecule. It is therefore important to determine whether the decomposition equilibrium is properly established within the cell. The geometry of the cell should be such that either the majority of molecules escape before decomposition or alternatively their residence time within the cell should be long enough to fully establish the decomposition equilibrium.[19]

In determining Kp for a reaction of the type $A \rightleftharpoons B+C$ it is necessary to know the pressure of each component. Now if component A has a vapour pressure at room temperature which is low compared with the pressure in the system, the vapour molecules will make a single pass through the ionization region and condense. If the component B is normally gaseous at room temperature its molecules will, on average, make several passes through the ionization region before being pumped away. Then if $P_A = P_B$ in the cell the partial pressure of B will be higher in the ion source. This effect was studied in a model 14 TOFMS using a known pressure of carbon dioxide from the decomposition of calcium carbonate and comparing it with a known pressure of silver. After eliminating the differences of γ_i and σ_i the ratio of sensitivities

$$\frac{S\,(\text{non-condensable gas})}{S\,(\text{condensable vapour})} = 6{\cdot}0. \tag{8}$$

This showed that there was a sixfold increase in the concentration of a gas over an easily condensed vapour such as silver in the ion source (type 14-107 with 7·5 l. sec baffle plate). There are, of course, a whole range of substances with intermediate vapour pressures at room temperature, but these are generally liquids, and in this type of work one is usually dealing with substances which are either solid or gaseous at room temperature. The value from eqn. (8) has therefore a general application for similar experimental conditions.

One possible source of ions or neutral decomposition products is the heating filaments and the effect of these may be studied separately. In the analysis of potassium sulphate vapour it was found that a small fraction of "spurious" K^+ ions did not originate in the ion source. The concentration of spurious K^+ ions could be measured with the source electron gun off. After a short period of heating the K_2SO_4 sample by radiation only at a constant temperature of 1370°K, the heating current was switched off and the spurious K^+ ion concentration measured over the

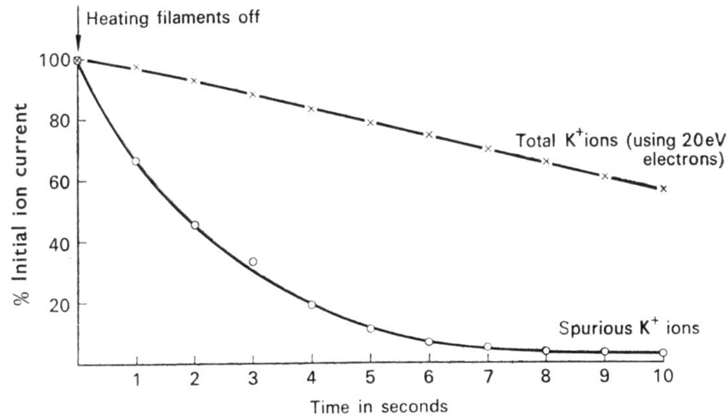

FIG. 6. The effect of sudden removal of cell heating of K^+ from K_2SO_4 vapour.

next 10 sec, during which time the ion current decreased rapidly to a very low value, Fig. 6. It was therefore concluded that these ions originated at or near the surface of the hot tungsten filaments. In the short time of 10 sec the thin heating filaments originally at around 3000°K cool very rapidly by radiation, whereas the effusion cell at 1370°K and with a much higher thermal capacity, cools only about 35°K in the same time.

The experiment was then repeated measuring the total K^+ ion concentration from K_2SO_4 vapour using 20 eV electrons. The total K^+ ion current was found to decrease much more slowly with time when the heating was switched off. It was thus shown that the majority of K^+ ions (99% in this case) were dependent on the temperature of the crucible rather than the heating filaments and that the mass spectrometer was sampling vapour effusing directly from the crucible orifice. This result was further confirmed by studying the effect on the ion current of tilting the crucible orifice with respect to the normal path of the molecular beam. The main K^+ ion current was entirely dependent on the alignment of the beam, whereas the spurious ion current remained constant.

THE VAPORIZATION OF INORGANIC MATERIALS BY SHOCK HEATING

The practical limit in studying high temperature reactions using conventional heating techniques such as electron bombardment, inductive or resistive heating is about 3000°K. The need to develop heat-resisting materials to protect space vehicles during

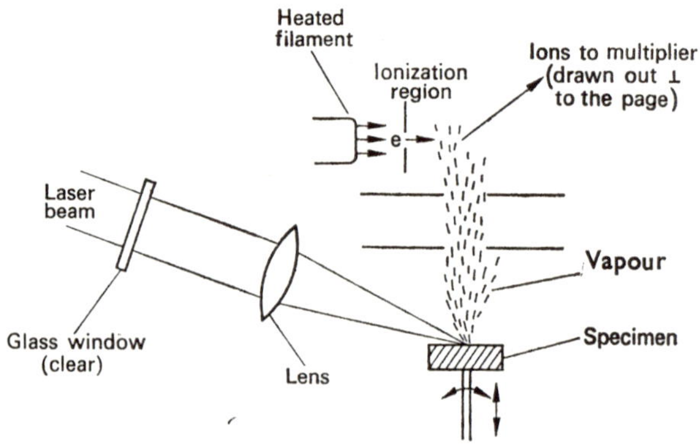

FIG. 7. Laser vaporization of carbon inside the TOF mass spectrometer (Zavitsanos).

re-entry of the atmosphere required thermodynamic data for temperatures in the region of 4000°K. These temperatures can be obtained using pulses of laser light focused on to the surface of the material. The advantage of laser heating is that only a small portion of the surface is vaporized with the rest acting as a container for the heated portion, thus eliminating reaction between container and specimen. A disadvantage is that the vapour may be in a state of thermal non-equilibrium and may also contain highly excited species such as ions.

P. D. Zavitsanos has studied the species liberated from laser heated carbon surfaces using a TOFMS.[20] The graphite samples were heated from the side at a low angle, Fig. 7. This reduced the path length of the laser beam through the vapour plume, which is always emitted normal to the sample surface. Experimental values for the rate of vaporization gave estimates of the heat of vaporization and the average molecular weight of the vapour. The mass spectrometer was used to determine the concentration of vapour phase species C_1–C_7 released during the laser heating time of 0·6 msec. A comparison with extrapolated data for carbon vapour in thermal equilibrium gave an estimated average surface temperature of 4080 ± 20°K.

Meyer and Ames[23] have obtained some qualitative results using a method of pulse heating thin wires of Ag, Zr, ZrO_2 and ZrO_xN_y to achieve Langmuir vaporization at temperatures in excess of 2000°K. This method again has the advantage of reducing the effects of material interaction. Temperatures were calculated on the basis of energy dissipation and were also measured using fast response optical pyrometry.

The reader is referred to the latest literature survey by Redman[21] in the field of mass spectrometric thermochemical technique which covers the period January 1957–April 1966. Current surveys from October 1966 onward appear in the *Mass Spectrometry Bulletin*.[22]

ACKNOWLEDGMENT

This paper is published by permission of the Central Electricity Generating Board.

REFERENCES

1. INGHRAM, M. G., and DROWART, J., *High Temperature Technology*, (McGraw-Hill, New York, 1960).
2. DROWART, J., *Condensation and Evaporation of Solids* (Gordon and Breach, New York, 1962).
3. GOLDFINGER, P., *Mass Spectrometry (NATO Symposium Glasgow)*, (Academic Press, London and New York, 1965).
4. WHITE, D., SOMMER, A., WALSH, P. N., and GOLDSTEIN, H. W., *Advances in Mass Spectrometry*, Vol. 2 (Pergamon Press, Oxford and London, 1963).
5. RAUH, E. G., SADLER, R. C., and THORN, R. J., *High Temperature Knudsen Effusion Cell Assembly*, Argonne National Laboratory Report ANL-6536, 1962.
6. PANISH, M. B., and REIF, L., *Rev. Sci. Instr.* **32,** 831 (1961).
7. KANT, A., *J. Chem. Phys.* **41,** 1872 (1964).

8. NESMEYANOV, A. N., *Vapour Pressure of the Chemical Elements* (Elsevier Publishing Co., Amsterdam, London, New York, 1963).
9. POTTIE, R. F., *J. Chem. Phys.* **44**, 916 (1966).
10. KENNARD, E. H., *Kinetic Theory of Gases with Introduction* to *Statistical Mechanics* (McGraw-Hill, New York, 1938).
11. ICZKOWSKI, R. P., MARGRAVE, J. L., and ROBINSON, S. M. *J. Phys. Chem.* **67**, 229 (1963).
12. OTVOS, J. W., and STEVENSON, D. P., *J. Am. Chem. Soc.* **78**, 546 (1956).
13. LAMPE, F. W., FRANKLIN, J. L., and FIELD, F. H., *J. Am. Chem. Soc.* **79**, 6129 (1957).
14. MANN, J. B., *J. Chem. Phys.* **46**, 1646 (1967).
15. STAFFORD, F. E., *J. Chem. Phys.* **45**, 859 (1966).
16. VERHAEGEN, G., STAFFORD, F. E., and DROWART, J., *J. Chem. Phys.* **40**, 1622 (1964).
17. KANT, A., *J. Chem. Phys.* **44**, 2450 (1966).
18. STANTON, H. E., CHUPKA, W. A., and INGHRAM, M. G., *Rev. Sci. Instr.* **27**, 109 (1956).
19. BOERBOOM, A. J. H., *Mass Spectrometry* (NATO Symposium Glasgow) (Academic Press, London and New York, 1964).
20. ZAVITSANOS, P. D., Technical Information Series Report R 67SD11, General Electric Space Sciences Laboratory, Missile and Space Division, Philadelphia, Pa., USA.
21. REDMAN, J. D., *A Literature Review of Mass Spectrometric Thermochemical Technique Applicable to the Analysis of Vapour Species over Solid Inorganic Materials*, Report ORNL-TM-989, Oak Ridge National Laboratory, Tenn.; Also Supplement I (1966).
22. *Mass Spectrometry Bulletin*, Mass Spectrometry Data Centre, AWRE Aldermaston, Berks., England.
23. MEYER, R. T., and AMES, L. L., to be published in *Applications of Mass Spectrometry in Inorganic Chemistry*, American Chemical Society Advances in Chemistry Series, Washington, DC 1967. Edited by J. L. Margrave.

SOME APPLICATIONS OF TIME-OF-FLIGHT MASS SPECTROMETRY TO INDUSTRIAL ANALYSIS

C. D'OYLY-WATKINS, D. E. HILLMAN and D. E. WINSOR

The Chemical Inspectorate, Royal Arsenal East, Woolwich, London

THE object of this paper is to discuss some of the applications for which a Bendix time-of-flight mass spectrometer has been employed in a laboratory which is called upon to analyse a wide range of manufactured products. These include paints, adhesives, polymers, textiles, oils, explosives and a variety of miscellaneous chemicals. Other problems encountered include the examination of degradation products from the above materials and investigation of changes in the atmospheres within sealed containers produced by deterioration of the enclosed stores.

Although purchased initially for use in conjunction with a gas chromatograph, as described by Gohlke[1] and by Ebert,[2] the instrument has also been found helpful in a number of other applications. In general the methods employed fall under four headings:

1. Qualitative identifications of pure materials eluted from a gas chromatograph.
2. Quantitative estimation using the mass spectrometer as a selective gas chromatographic detector.
3. Examination of volatile materials and gaseous mixtures by mass spectrometry alone.
4. Examination of non-volatile materials using the Bendix 843A direct inlet system.

QUALITATIVE IDENTIFICATION OF MATERIALS ELUTED FROM A GAS CHROMATOGRAPH

Gas chromatography, although an excellent technique for the separation of organic mixtures, can provide ambiguous information as to the identity of the separated fractions. The relative retention time of a material on a specific column gives information concerning the vapour pressure of the material, at the column temperature, above its solution in the liquid used as the stationary phase. Alteration of the stationary phase or of the column temperature will modify this vapour pressure and agreement of relative retention time between a known and an unknown material on a number of columns can be taken as an indication of the identity of these materials. This evidence is not absolutely conclusive and its collection is time-consuming if a number of columns need to be employed. For these reasons other methods, such as collection of emergent fractions for examination by chemical or infrared spectrophotometric means, are often employed for confirmation.

Mass spectrometry can produce a "finger print" of the compound and possesses a sensitivity such that it can deal with amounts of material which are present in fractions eluted from analytical size samples. It therefore provides an ideal confirmatory technique which can be coupled directly to the chromatograph without the need for intermediate collection of fractions.

Until the advent of interrupted elution chromatography a fast scanning speed was the first essential requirement in a mass spectrometer employed for this purpose. The normal times of 10–20 min required to scan a spectrum by variation of magnetic field strength in conventional instruments are obviously of little use since a large number of fractions can emerge from the chromatograph during this time. Even scan speeds of one-fifth of a second, achieved over a limited mass range by variation of the electrical field, can sometimes be insufficiently fast. We have found, by measurement on the ascending slopes of early sharp peaks, that the concentration of sample in the carrier gas can double in one tenth of a second. Usable spectra could be obtained with scanning speeds of this order provided that they were taken at the time of the peak maximum where concentration of the sample is changing much more slowly. Unfortunately this time cannot be determined by examination of the trace produced from the gas chromatograph detector because of the lag introduced by the length of piping connecting the chromatograph and mass spectrometer. With the time-of flight instrument the scan time of 100 μsec is so short that virtually no change in sample concentration occurs even with the sharpest peaks. The oscilloscope trace provides in effect, a continuous monitor of the spectrum of the gas leaving the chromatograph. When a sample component is eluted, all the mass peaks rise and fall simultaneously on the screen and a photograph can be taken at any time. Two photographs, one taken when a fraction starts to elute and a second just before it finishes, can provide spectra of pure components incompletely resolved by the chromatograph.

In order to transfer the eluted fractions directly into the mass spectrograph we have tried a number of connecting systems including the simple method of bleeding a small fraction of the column effluent through a capillary restriction straight into the spectrometer source. Our best results were produced by the use of an intermediate chamber, continuously pumped by an auxiliary system, in which a pressure of some 0·2 torr could be maintained. This system provides a rapid rise and fall of sample concentration within the mass spectrometer source and reduces tailing to a minimum. No attempt has been made to concentrate the sample in the carrier gas by methods such as selective diffusion through Teflon. This is partly because we have not found it necessary with the sensitivity available with the Bendix instrument and the size of sample which we normally use, but mainly in order to avoid, as far as possible, selective removal of sample components which could affect the quantitative use of the system.

The sample system which we are at present using is illustrated in Fig. 1. The gas flow is divided at the end of the column so that part goes to a flame ionization detector and the remainder passes through a stainless steel capillary of 0·1 mm internal diameter to the sample inlet valve V_1. The capillary can be heated by the passage of an electric current through its walls. Since both ends of the capillary are connected to points on the apparatus which are at earth potential, current is fed to its centre point and travels through both halves in parallel. The remainder of the inlet system is heated by an electrical tape. Pressure in the sample system is maintained at about 0·2 torr by adjustment of valve

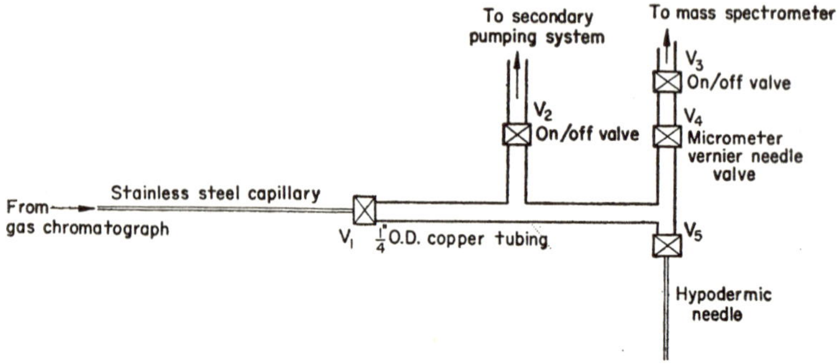

FIG. 1. Diagram of sample inlet system.

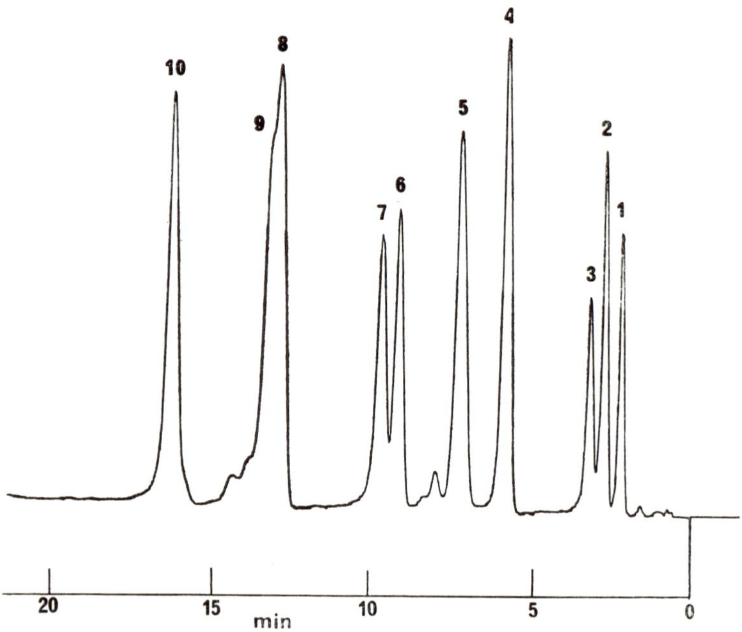

FIG. 2. Chromatogram of a glycol ether mixture on a 5 ft×4 mm column packed with 10% Carbowax 20M on Gas Chrom Z. Temperature programmed 100–225°C at 6°C per min. Detector flame ionization. Peaks: 1, propylene glycol monomethyl ether; 2, ethylene glycol monomethyl ether; 3, ethylene glycol monoethyl ether; 4, ethylene glycol monobutyl ether; 5, dipropylene glycol monomethyl ether; 6, diethylene glycol monomethyl ether; 7, diethylene glycol monoethyl ether; 8, diethylene glycol monobutyl ether; 9, tripropylene glycol monomethyl ether; 10, triethylene glycol monomethyl ether.

V_1 and by pumping through V_2 with the 843A direct inlet system to which it is connected by a T-joint. The sample is admitted to the mass spectrometer through a micrometer valve V_4 which is opened to the maximum orifice possible without undue loss of vacuum in the spectrometer. Valve V_3 allows the system to be isolated without interference with the micrometer valve setting. The piping between valve V_3 and the mass spectrometer source is made as short and wide as possible to avoid unnecessary loss of pumping speed.

The system has been found to work well with samples of boiling point up to about 300°C. Sizes of about 0·2 μl are normally employed for samples but 0·001 μl of n-decane injected on to the column has been detected on the oscilloscope. For fractions other than those giving early very sharp peaks, a number of output scanners, covering separate parts of the spectrum simultaneously, enable the whole spectrum to be produced on a fast multichannel recorder with considerable increase in sensitivity. Figure 2 shows the chromatogram obtained by injecting a mixture of 10 glycol ethers on to the chromatograph and Fig. 3 the spectrum produced by each component. Compounds of this type are found in a wide range of materials, such as hydraulic and de-icing fluids. Difficulty can be experienced in the gas chromatographic analysis of such mixtures because of the wide boiling range of the constituents. Programming of the column temperature during the run is usually carried out in order to shorten the analysis time and to sharpen the later peaks. Under programme conditions normal isothermal relative retention times cannot be used and actual retention times often vary due to slight differences in the column temperature at the start of separate runs. Direct comparison of the photographs of mass spectra obtained from known materials with those appearing on the oscilloscope during a sample run has enabled us to identify easily the components of mixtures of this type. The monomer and polymers of the same ether usually give very similar mass spectra but such compounds are widely separated on the chromatograph and confusion between them is unlikely.

For the purpose of immediate comparison with spectra of sample fractions, photographs of the oscilloscope screen are the ideal method of producing reference data. Unfortunately, with some compounds, adjustment of the time base speed to accommodate all the masses required on the trace results in severe cramping of the peaks with consequent difficulty in identification of individual masses. In an endeavour to overcome this difficulty we sought for some method of expanding the scale of the trace. The oscilloscope normally used with the Bendix mass spectrometer possesses dual trace characteristics by means of which two different input signals can be displayed on alternate sweeps. Connection of the mass spectrometer output to both oscillo-

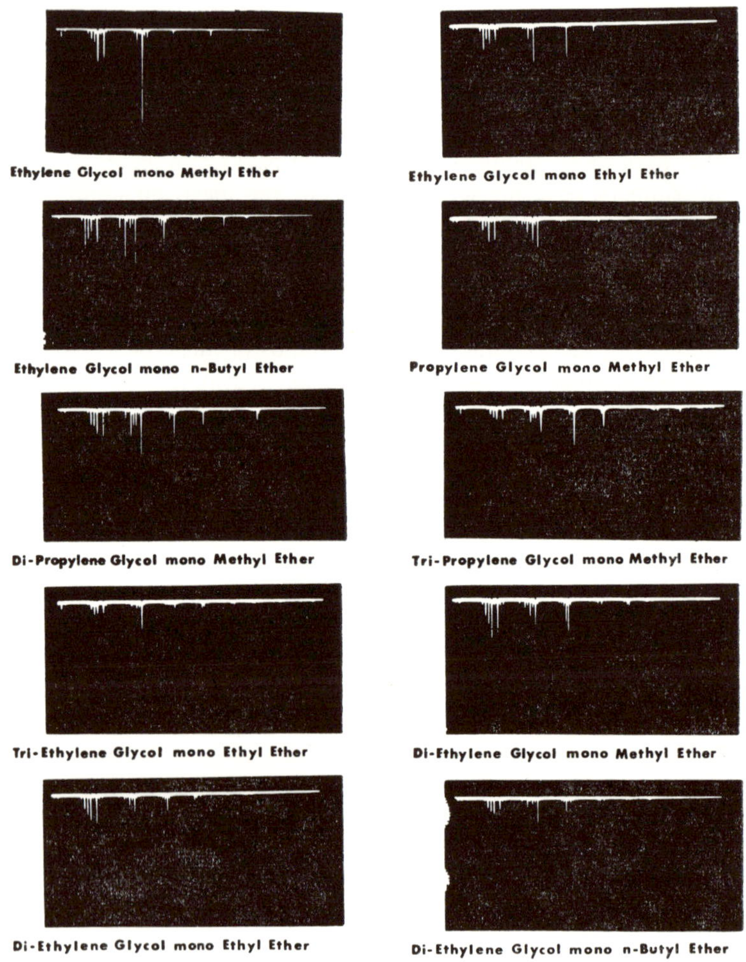

FIG. 3. Glycol mono ethers. Mass spectra of components eluted during the production of the chromatogram shown in Fig. 2.

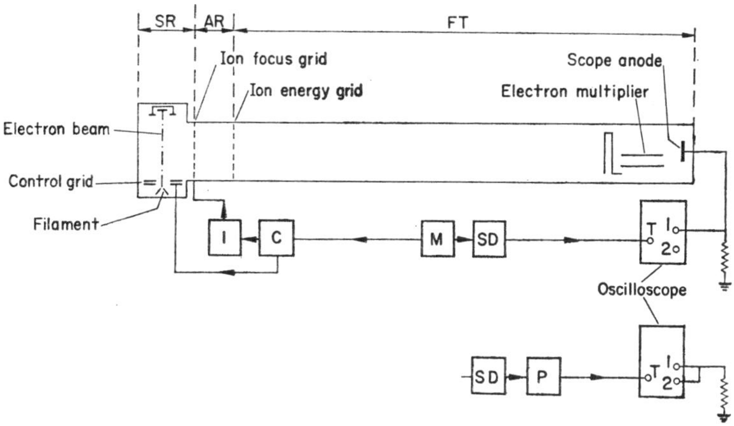

FIG. 4. Schematic diagram showing modifications for the production of dual trace display. *SR*, source region. *AR*, accelerating region. *FT*, flight tube. *I*, ion pulse generator. *C*, control grid pulse generator. *M*, master pulse generator. *SD*, scope delay pulse generator. *P*, pulse generator (Solartron Type GO1101-2). *T*, oscilloscope trigger input. 1, 2, oscilloscope signal inputs.

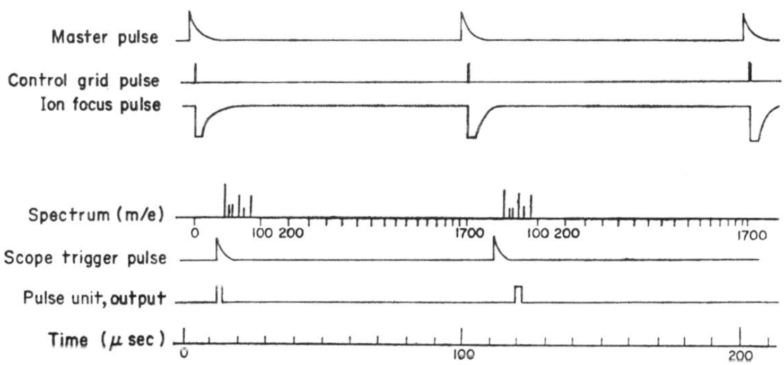

FIG. 5. Diagram showing time relationship of mass spectrometer pulses.

scope inputs produces two identical displays one of which can be inverted relative to the other. If the delay time of the trigger pulse with respect to the start of the spectrum can be made greater on alternate cycles, one of the traces will display a later part of the spectrum than that shown by the other. It was found that this could be accomplished quite easily by inserting a Solartron pulse generator, Type GO 1101-2, in the lead between the mass spectro-

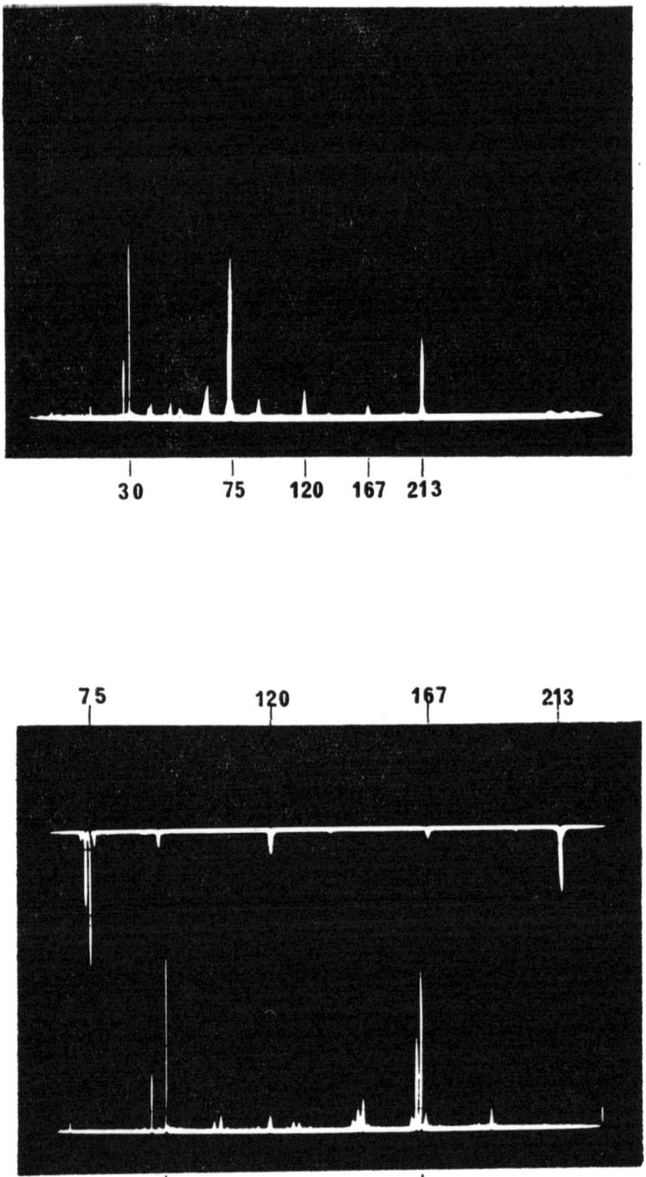

FIG. 6. Photographs showing comparison between (a) single, and (b) dual trace presentation of the spectrum of 2,4,6-trinitrobenzene.

meter trigger output and the oscilloscope trigger input, as shown in Fig. 4. This generator can be made to give two pulses on each cycle, one immediately after the start of the cycle and the other which can be delayed by anything up to the periodic time after it. If the generator is set to run with a periodic time of 200 μsec and synchronized with the spectrometer output it will fire only on alternate cycles of the spectrometer. Adjustment of the delay to a time greater than 100 μsec results in a display of two separate parts of the spectrum on the oscilloscope. The time relationship of the various pulses involved is shown in Fig. 5 and a comparison spectra of the same material using single and double trace presentation is shown in Fig. 6.

QUANTITATIVE ESTIMATION BY COMBINED GAS CHROMATOGRAPHY AND MASS SPECTROMETRY

Quantitative analysis by gas chromatography is often rendered difficult by incomplete separation of components. This presents problems in assessing the contribution of the height of one peak to that of the other unresolved peak or of dividing the combined area when this method is employed. Use of a selective detector is possible in some cases, as in the estimation of pesticides by electron capture. Here, peaks which would be produced by the solvents present, and which would normally obscure that of the active material, are eliminated. In a paper on the combined techniques, Henneburg and Schomburg[(3)] described the use of a mass spectrometer running at a fixed mass number to produce a chromatogram showing peaks only for those components whose spectra contained ions of the chosen mass. Repeat chromatographic runs using different spectrometer settings enabled a considerable amount of qualitative information concerning the components of the mixture to be obtained. This method could be used even with spectrometers of the slowest scanning speed.

The idea of using a mass spectrometer as a selective detector for the quantitative estimation of partially resolved fractions would be inviting were it not for one main difficulty. Reproducibility in the size of the minute samples introduced on to a gas chromatograph is notoriously bad and accurate measurement of the actual size even worse. For this reason it is usual to measure ratios of peak heights or areas which are independent of sample size. For absolute determinations a known amount of an internal standard is added to a known amount of the bulk sample before use. Most mass spectrometers are capable of monitoring only one ion abundance at a time, and repeat chromatographic runs would be necessary for each component and probably for an internal standard. The time-of-flight mass spectrometer does not suffer from this disadvantage since, by the use of several output scanners, all the chromatograms required can be produced on a multi-channel recorder from a single run.

The use of this instrument should make the quantitative estimation of two chromatographically unresolved materials a comparatively easy matter. This is particularly the case if the spectrum of each component contains a mass peak of reasonable size which is absent in the spectrum of the other material. The mass spectrograph then behaves as if it were two chromatographic detectors each sensitive to one of the components only. A calibration run using a mixture of equal parts of the two substances enables the relative sensitiveness of the detectors to be calculated and the relative amounts of each of these in a sample can then be calculated from measured peak heights or areas on the two chromatograms produced from a single sample run. If absolute figures are required these can be obtained by adding a known amount of an internal standard to both calibration and sample mixtures. Either of the two masses used for the sample can be used to produce the internal standard peak provided that one, or both are present in its spectrum. Failing this a third mass could be used to give a third chromatogram containing the standard peak.

In the event of it proving impossible to find mass peaks for each component which are essentially absent in the spectrum of the other component any pair of masses which produce different ion abundances can be used. It is then necessary to perform two additional chromatographic runs, one for each of the pure materials. From the four pairs of chromatograms obtained, four ratios of peak heights or areas can be measured. Over the linear range of the instrument the peak area and in some cases the peak height will be equal to the concentration of the material injected multiplied by a sensitivity factor for the particular material at the mass number employed.

Then, for two masses x and y, if a = peak size produced by unit quantity of substance A at mass indicated by subscript; b = peak size produced by unit quantity of substance B at mass indicated by subscript; C = amount of substance indicated by subscript injected to produce chromatogram indicated by superscript (A = pure A. B = pure B. 50 = 50/50 mixture. S = sample); p = size of measured peak produced at mass indicated by subscript by injection of material indicated by superscript; r = ratio calculated from measured peak sizes from injection of material indicated by subscript,

$$r_A = \frac{p_y^A}{p_x^A} = \frac{C_A^A\, a_y}{C_A^A\, a_x}.$$

Taking ratios therefore eliminates the concentration and

$$r_A = \frac{a_y}{a_x}. \tag{1}$$

Similarly,

$$r_B = \frac{p_x^B}{p_y^B} = \frac{b_x}{b_y}. \tag{2}$$

For the 50/50 mixture since ion abundances are additive:

$$r_{50} = \frac{p_x^{50}}{p_y^{50}} = \frac{C_A^{50}a_x + C_B^{50}b_x}{C_A^{50}a_y + C_B^{50}b_y}$$

and, since

$$C_A^{50} = C_B^{50},$$

$$r_{50} = \frac{a_x + b_x}{a_y + b_y}. \tag{3}$$

For the sample chromatograms

$$r_s = \frac{p_x^s}{p_y^s} = \frac{C_A^s a_x + C_B^s b_x}{C_A^s a_y + C_B^s b_y}; \tag{4}$$

rearranging gives the required ratio,

$$\frac{C_A^s}{C_B^s} = \frac{b_x - r_s b_y}{r_s a_y - a_x}$$

and by substitution throughout eqns. (1.4),

$$\frac{C_A^s}{C_B^s} = \frac{r_B - r_A r_B r_{50} + r_A r_{50} r_s - r_s}{r_B - r_A r_B r_s + r_A r_{50} r_s - r_{50}}.$$

In the case where there is no cross contribution from one component to the peak used for the other component, $r_A = r_B = 0$, and the equation reduces to

$$\frac{C_A^s}{C_B^s} = \frac{r_s}{r_{50}}.$$

In order to assess the quantitative accuracy obtainable when using a mass spectrometer as a selective detector, 0·5 μl of a mixture of benzene and toluene was injected on to the chromatograph. This mixture was chosen since it can be easily separated chromatographically and comparative determinations could be carried out by the simultaneous use of both detection systems. The spectra which are given by each of these materials are shown in Fig. 7. Mass 91 was chosen for the detection of toluene as it is the base peak of this spectrum and completely absent in that of benzene. Mass 78 was chosen for the detection of benzene. In this case there is a slight contribution at mass 78 in the toluene spectrum, but it is in fact small enough to be negligible unless the toluene content of the mixture is very high. The three chromatograms obtained from the flame ionization detector and from the output scanners are also shown in the figure. Peak heights were measured and the ratio of that of the benzene peak to that of the toluene peak calculated in each case. These ratios were not equal because of differences in sensitivity of the individual detection systems. A second mixture of benzene and toluene of slightly different composition was now run and the ratios similarly calculated. Division of the first ratio by the second for each instrument eliminated the sensitivity factors and gave the ratio of benzene in mixture 1 to that in mixture 2. As can be seen in Fig. 7, excellent agreement was obtained between the two methods.

A second experiment, using a mixture of diethyl ketone and methyl n-propyl ketone is illustrated in Fig. 8. These compounds have identical molecular weights, similar boiling points and are of the same chemical type. They are therefore difficult to separate chromatographically and impossible on the columns we have currently available. In this case, mass 57 was selected for the estimation of diethyl ketone and mass 43 for methyl n-propyl ketone. Careful examination of the individual peaks (b) and (c) show, in fact, a slight separation of the two components, methyl n-propyl ketone being eluted fractionally earlier than diethyl ketone. This was to be expected from the slight difference in boiling points but was insufficient for resolution with the normal detector which produced only a single peak (a).

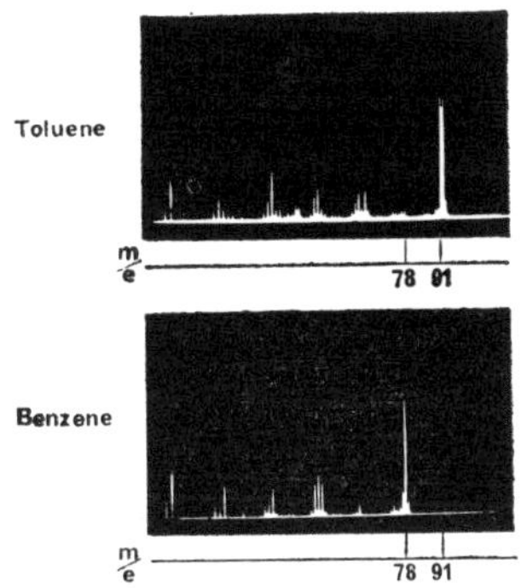

(a)

(b)

(c)

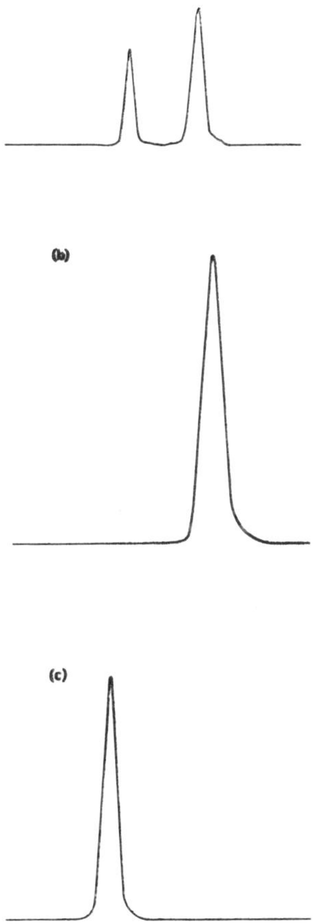

0·5 μl of a benzene/toluene mixture injected onto an apiezon L column at 125° C.

Three recordings taken simultaneously show:—

(a) Response of flame ionisation detector to both components.

(b) Response of one output scanner set to monitor mass/charge 91 only–indicating in this case the toluene component.

(c) Response of a second output scanner set to monitor mass/charge 78 only – indicating in this case the benzene component.

RESULTS OF QUANTITATIVE RUNS

	Ratio of Benzene pk. ht./Toluene pk. ht.	
	G/C	M/S
Mixture A	{0·848, 0·851, 0·842} 0·848	{2·23, 2·19, 2·08} 2·17
Mixture B	{0·761, 0·757, 0·747} 0·755	{1·95, 1·88, 1·88} 1·90
Ratio $\frac{\text{Mixt. A}}{\text{Mixt. B}}$	1·12	1·13
Factual Ratio $\frac{\text{Mixt. A}}{\text{Mixt. B}}$ 1·1		

FIG. 7. Results obtained from quantitative runs of benzene–toluene mixtures using the mass spectrograph as a multiple selective chromatographic detector.

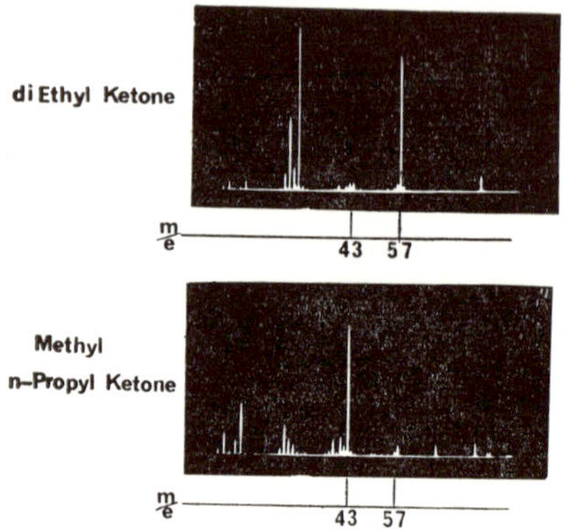

(a)
(b)
(c)

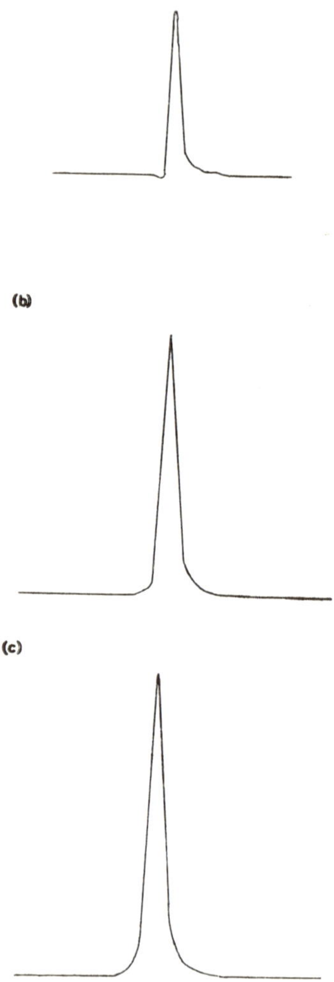

0·5μl of diethyl ketone/methy n-propyl ketone mixture injected onto an apiezon L column at 125°C.

Three recordings taken simultaneously show:–

(a) Response of flame ionisation detector; components unresolved by chromatograph.

(b) Response of one output scanner set to monitor mass/charge 57 only – indicating in this case the diethyl ketone component.

(c) Response of another output scanner set to monitor mass/charge 43 only – indicating in this case the methyl n-propyl ketone component.

RESULTS OF QUANTITATIVE RUNS

	PEAK HEIGHTS		RATIO
	Me n-Pr. K.	di. Eth. K.	
MIXTURE A.	54·0	38·5	1·40
51·0% MenPr. K	54·0	38·5	1·40
49·0% di Eth. K	50·0	36·0	1·39
	53·0	38·0	1·39
	49·0	35·5	1·39
	52·0	37·0	1·41
MIXTURE B.	47·0	24·5	1·92
	52·0	26·5	1·96
	58·0	30·0	1·93
	40·0	21·0	1·91
	29·0	15·5	1·87
	61·0	31·0	1·97
%Experimental	57·0	43·0	
%Theoretical	58·5	41·5	

FIG. 8. Quantitative analysis of a diethyl ketone–methyl n-propyl ketone mixture by combined gas chromatography and mass spectrometry.

Using ratios obtained from the chromatograms and the known composition of mixture *A* (49% diethyl ketone and 51% methyl n-butyl ketone) the composition of mixture *B* was calculated by standard internal normalization techniques. The results obtained show good agreement with the actual composition.

There is in this case an appreciable contribution of each of the ions chosen in the spectrum of the other compound and it is possible that even better agreement would have been obtained had this been taken into account. On the other hand, the contribution at mass 43 is due to $-C_3H_7^+$ ions which should be produced only by the methyl n-propyl ketone and the contribution at mass 57 to $C_2H_5CO^{\pm}$ ions which are unlikely to be given by this compound. It is probable therefore that the cross contributions are due to contamination of each substance with a trace of the other. In this event the experimental results may well be the better estimate of the true composition of the mixture.

From the results given it can be seen that, provided the interconnecting system is designed to avoid, as far as possible, mass discrimination effects, particularly those which vary with component concentration, the time-of-flight mass spectrometer provides a valuable means for the estimation of unseparated components of a chromatographed mixture.

An additional bonus is obtained with this method because the gain and attenuation of the output scanners can be separately adjusted, for mixtures where cross contributions can be neglected, to give equal sensitivities for each component. Choice of a sample size to give full scale deflection with a pure sample of one of the constituents of the mixture enables an approximate percentage composition to be obtained by directly reading the percentage full scale height for each peak.

EXAMINATION OF VOLATILE MATERIALS AND GASEOUS MIXTURE

The sample system (Fig. 1) provides an additional method of sample introduction through a hypodermic needle and valve V_5. Samples of volatile materials contained in small glass tubes closed with serum caps can be conveniently examined by this system. The needle is inserted through the rubber cap with the tube immersed in liquid nitrogen so that the sample is frozen. Any air in the tube can then be pumped away and the sample subsequently vaporized, by heating if necessary. Alternatively small amounts of samples can be injected, by means of a hypodermic syringe, into a pre-evacuated tube.

Stable spectra can be maintained for long periods in this way, thus enabling more detailed inspection of the spectrum to take place and slower recording speeds to be employed than is possible with a coupled chromatograph. Setting up of the output scanner gates for quantitative analysis of a chromatographed mixture is most easily and precisely accomplished on such a stable spectrum produced from a small quantity of the mixture.

Our major use of this system has been for the examination of atmospheres within sealed vessels. Climatic trials of components or materials in closed containers often result in changes in composition of the atmosphere within the container. This can be either depletion of one constituent, such as oxygen, or the appearance of gaseous decomposition products from the contents. The containers are provided with openings sealed by serum caps so that sampling can be carried out by insertion of a hypodermic needle.

Without a molecular inlet system capable of maintaining a known volume of gas at a controlled temperature and pressure and providing molecular flow into the source, accurate analysis of gaseous mixtures is very difficult. Nevertheless, we have been able to obtain approximate but useful analyses from samples which have been subjected to climatic trials.

In practice the atmosphere from the can is pumped slowly but continuously through the sampling system whilst allowing the amount required to produce a spectrum to leak through valve V_4 into the source. Continuous pumping has been found to reduce build up of concentration of compounds desorbed from the walls of the sampling system and of atmospheric components entering through minute leaks.

Absolute determination of the sensitivities of the spectrometer for the various ingredients is not possible in the absence of more precise sample metering facilities. We therefore use the argon in the atmosphere as an internal standard on the assumption that this will not take part in any reaction and will be identical in concentration to that found in normal air. Variations in pressure from sample to sample are corrected by measuring the ratio of the heights of peaks used for the determination of each ingredient to that of mass 40, assumed to be due entirely to argon.

A calibration curve obtained by adding known amounts of

carbon dioxide to air is shown in Fig. 9 and indicates the order of accuracy which can be obtained. The large value of 0·3% CO_2 for air alone (instead of the known 0·03%) which can be obtained by extrapolation of the line to the *X*-axis, is due to the conversion of some of the oxygen present in the samples to carbon dioxide on the source filament. A correction for this, calculated from the measured oxygen content, can be applied.

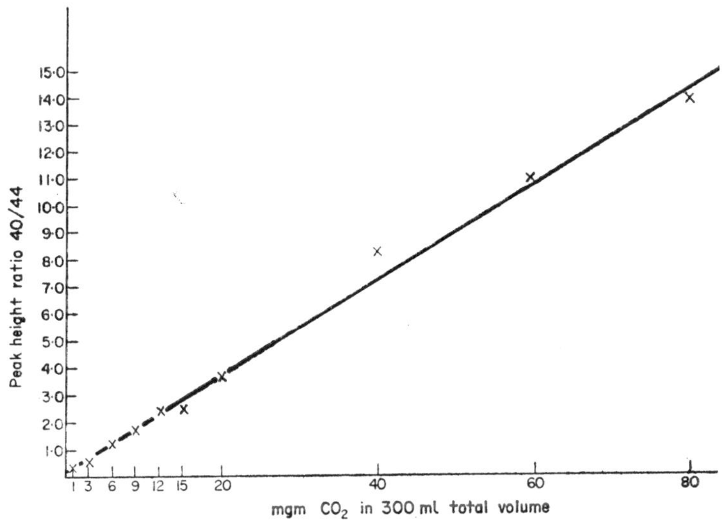

FIG. 9. Calibration curve for the determination of carbon dioxide in atmospheres.

For the examination of atmospheres within small glass containers, such as radio valves, the device shown in Figs. 10 and 11 has been of valuc. This consists of a metal plate to which is fitted an Edwards "break seal" valve. A tightly fitting steel rod pointed at one end, passes through the valve the O-ring seal of which prevents ingress of air into the assembly. This plate is bolted on to a flanged metal container, of suitable size to contain the sample, with an interposed rubber gasket. A side arm is connected to valve V_5 in place of the hypodermic needle and after pumping out the atmosphere within the container a sharp tap on the end of the steel rod fractures the glass wall of the sample liberating the contained gas.

EXAMINATION OF NON-VOLATILE MATERIALS

The Bendix 843A direct inlet system is designed to permit small samples of solids or non-volatile liquids contained in small silica tubes to be introduced directly into the source chamber of the spectrometer. The sample tube sits upon a thermocouple mounted on the end of an accurately machined metal probe. Surrounding the tube is a small helical heating coil, the connections for this and for the thermocouple passing down through the length of the probe through vacuum tight channels. In use the probe is inserted into a sleeve, mounted below the spectrometer source and fitted with O rings to provide a vacuum seal. Air is pumped out from above the probe by an auxiliary pumping system after which the probe can be pushed further into the instrument, through a vacuum valve. In its final position the open end of the sample crucible is positioned within a millimetre or so of the source electron beam.

Samples inserted in this way can be volatilized by passage of a current through the heating coil. The molecules of vapour produced pass straight into the ionization region of the spectrograph without prior collision with any walls. Difficulties due to adsorption and condensation effects normally encountered when using a standard vapour inlet system are avoided and excellent spectra can be obtained from a very wide range of materials.

We have found this inlet system of great value for the rapid identification of polymeric materials. A small chip from a plastic sample can be inserted in the crucible and heated until a spectrum is obtained. Although usually produced by a combination of pyrolysis and electron bombardment the ion abundance patterns produced have, with all materials we have so far examined, been characteristic of the species present. Samples of the same material from different sources have shown good reproducibility and distinctions between different species have been quite clear. Figure 12 shows the spectra obtained with three types of nylon. Differences in the patterns obtained are clear and can to some extent be related to the chain lengths of the monomeric ingredients used to produce the plastic.

A final application of the instrument, which also employs the direct inlet system, has been for the examination of explosive materials. Compounds of this type, particularly those which

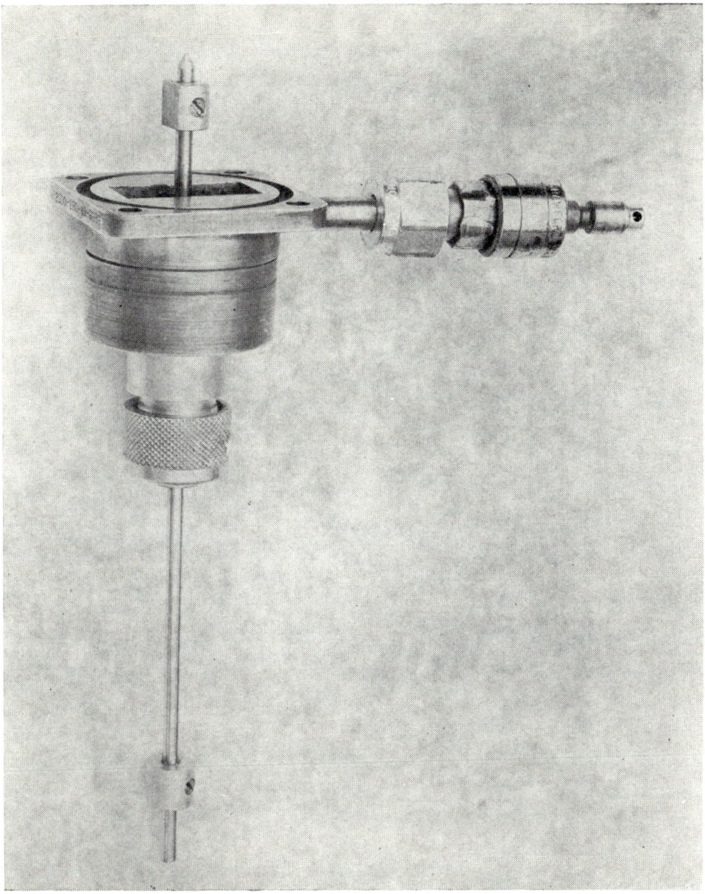

FIG. 10. Device for sampling atmospheres within small glass container.

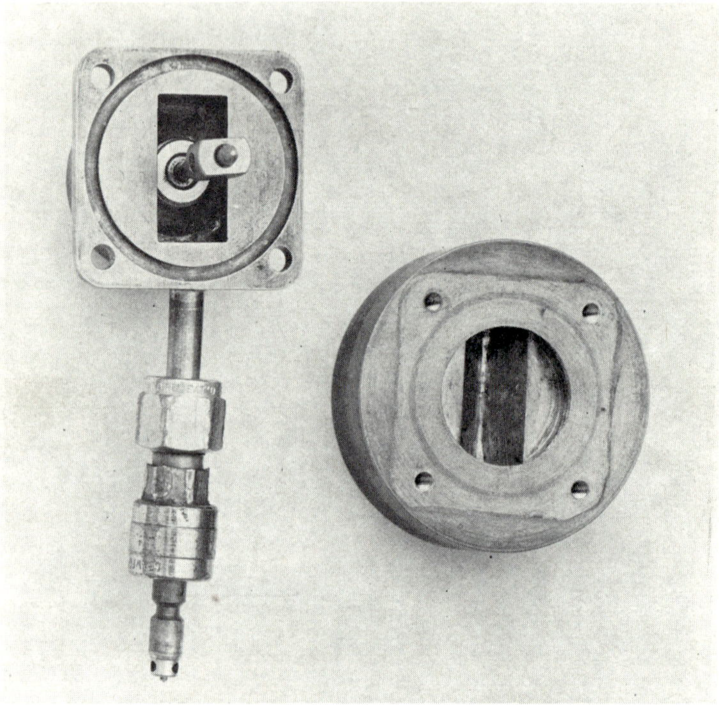

FIG. 11. Device for sampling atmospheres within small glass container.

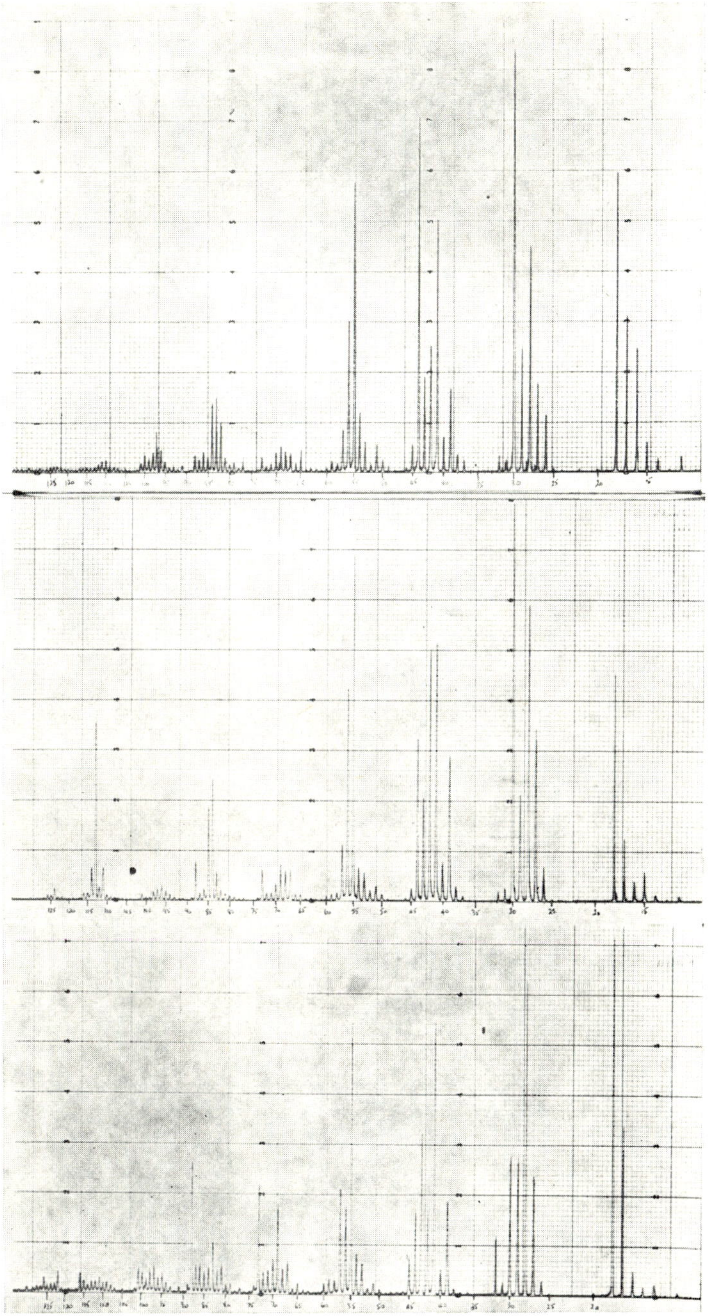

FIG. 12. Mass spectra of three nylons produced with the use of the Bendix direct inlet system *(top)* Nylon 6.6, *(middle)* Nylon 6, *(bottom)* Nylon 11.

contain nitrated aromatic rings, have been found to give large parent peaks. Other peaks in the high mass end of the spectrum are usually few and at easily deducible mass difference from the parent. The presence of a peak at an unexpected mass number can be clearly seen and can frequently be identified as belonging to the spectrum of another compound of this class. Examination by this method can therefore provide a rapid method for the detection and identification of impurities in explosive materials.

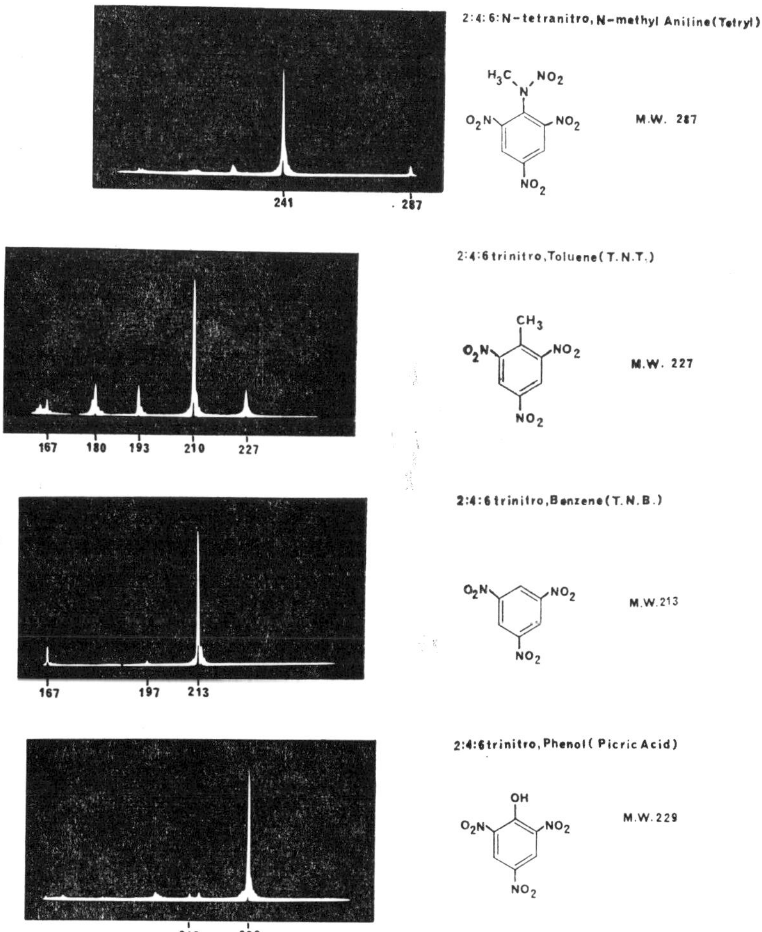

FIG. 13. Photographs of high mass peaks from spectra of various explosives.

Figure 13 shows the spectra of four compounds of this type where the characteristic fragmentation pattern can be clearly seen. The largest fragment ion in the spectra is usually produced by loss of an oxygen atom from one of the nitro groups and appears as mass 213 in the spectrum of picric acid and at mass 197 in that of 2,4,6-trinitrobenzene. 2,4,6-trinitrotoluene produces an unusual ion in large abundance at mass 210, corresponding to a loss of —OH. A similar ion has been found in the spectrum of 2,4-dinitroaniline by Spiteller[4] who suggested that a six-membered ring is first formed by internal hydrogen bonding between one of the amino hydrogens and an oxygen atom in the adjacent nitro group. Electron shifts around the ring then result in the elimination of —OH. This peak is therefore diagnostic of the presence of an ortho nitro group in nitro toluenes. In this case the second ortho group in the trinitro compound produces a further loss of —OH to produce the peak at mass 193. In 2,4,6-*N*-tetra nitro *N*-methyl aniline the interpolation of a nitrogen atom between the methyl group and the ring prevents this mechanism from operating and the highest fragment at mass 241 is due to the loss of a nitro group, presumably from the side chain.

CONCLUSION

In this paper we have been able to describe only a few of the many analytical applications for which this very versatile technique can be used. Combination with gas chromatography extends greatly the complexity of mixtures which can be examined and obviously has greater advantages for the analysis of volatile mixtures than are provided by the use of either technique alone. Similar advantages for higher boiling materials can undoubtedly be obtained by combination of mass spectrometry with other techniques such as thin layer or gel permeation chromatography.

REFERENCES

1. Gohlke, R. S., *Anal. Chem.* **31** (4) 535–41 (1959).
2. Ebert, A. A., Jr., *Anal. Chem.* **33** (13) 1865–70 (1961).
3. Henneburg, D., and Schomburg, G., *Gas Chromatography* (Edited by van Swaay, M., Butterworths, London, 1963), pp. 191–202.
4. Spiteller, G., Monatsh **92,** 1147 (1961).

BIBLIOGRAPHY

W. K. Joy

The British Coal Utilization Research Association, Leatherhead, Surrey

INTRODUCTION

Every effort has been made to include all published papers which are readily available at least up to the end of March 1968. Papers presented at conferences whose proceedings are restricted and do not constitute publication have been excluded. Publications subsequent to March 1968 can be readily traced by means of the abstracting journal *Mass Spectrometry Bulletin* published by the Mass Spectrometry Data Centre, AWRE, Aldermaston, Berks., England.

The Bibliography is divided according to application. The references have been placed in the most appropriate section for their content and are in chronological order in each section. In some cases a paper could equally well be placed in another category (e.g. qualitative analysis and chromatographic analysis). Care should, therefore, be taken when using the Bibliography since each paper is only quoted once.

My thanks are due to Dr. J. H. Lippiatt and Messrs. S. Howarth and G. B. Ward at the University of Salford for assistance in checking the references.

REVIEWS

Time-of-flight mass spectrometry. Harrington, D. B., *Encyclopedia of Spectroscopy* (Ed. Clark, C. F.), Reinhold, NY (1960), 628.

The time-of-flight mass spectrometer—its applications. Tomita, H., *J. Chem. Soc. Japan* **67,** 1769 (1964).

Le spectromètre de masse à temps de vol et ses applications. Couchet, G., and Zvenigorosky, A., *Rev. Quest. Sci.* **26,** 167 and 334 (1965).

Recent advances in time-of-flight mass spectrometry. Damoth, D. C., *Adv. in Anal. Chem. and Instrumentation*, Wiley, NY **4,** 371 (1965).

Advances in time-of-flight mass spectrometry. Damoth, D. C., *Mass Spectrometry* (Ed. Reed, R. I.,), Academic Press, London (1965), 61.

Review of time-of-flight mass spectrometry. Thirion, B., *Méthodes Phys. Anal.* 1966, 62.

Characteristics of time-of-flight mass spectrometry. Hikita, T., Ishihara, F., Sueyasu, S., and Tsuchiya, S., *Mass Spec. Japan* **14,** 157 (1966).

Time-of-flight mass spectrometry. Joy, W. K., *British Coal Utilization Research Association Monthly Bulletin* **31,** 581 (1967).

The time-of-flight mass spectrometer. Price, D., *Chem. Brit.*, 4. June 1968.

Abstracts and Bibliographies on the Bendix Time-of-flight Mass Spectrometer, published by the Cincinnati Division of the Bendix Corporation, 3625 Hauck Road, Cincinnati, Ohio 45211, USA.

INSTRUMENTATION

New time-of-flight mass spectrometer. Katzenstein, H. S., and Friedland, S. S., *Rev. Sci. Instr.* **26,** 324 (1955).

Time-of-flight mass spectrometer with improved resolution. Wiley, W. C., and McLaren, I. H., *Rev. Sci. Instr.* **26,** 1150 (1955).

Method of analysis of an ion beam. Heym, A., Joseph, C., and Loude, J. F., *Helv. Phys. Acta* **34,** 444 (1961).

Improved sampling and recording systems in gas chromatography—time-of-flight mass spectrometry. Ebert, A. A., Jr., *Anal. Chem.* **33,** 1865 (1961).

Resistance strip magnetic electron multiplier. Goodrich, G. W., and Wiley, W. C., *Rev. Sci. Instr.* **32,** 846 (1961).

Experimental methods for studying free radicals using a time of flight mass spectrometer. Harrington, D. B., *Fifth International Symposium on Free Radicals, Uppsala* (1961).

A modified time-of-flight mass spectrometer for studying ion–molecule or neutral particle–molecule interactions. Lehrle, R. S., Robb, J. C., and Thomas, D. W., *J. Sci. Inst.* **39,** 458 (1962).

High temperature Knudsen effusion cell assembly. Rauh, E. G., Sadler, R. C., and Thorn, R. J., *Argonne Nat. Lab.*, ANL Report 6536 (1962).

Continuous ion source for a time of flight mass spectrometer. Studier, M. H., *Rev. Sci. Inst.* **34,** 1367 (1963).

A high temperature inlet manifold for coupling a gas chromatograph to the time of flight mass spectrometer. Miller, D. O., *Anal. Chem.* **35**, 2033 (1963).

Calibration of a flyable mass spectrometer for nitrogen and oxygen atoms. Cohen, H. A., Morgan, J. E., Narcissi, R. S., and Schiff, H. I., (McGill Univ., Montreal, Canada), AFCRL-63-437 (1963).

Time-of-flight mass spectrometer with sampling conversion of output signal. Anufriev, G. S., and Mamyrin, B. A , *Prilsurg i Tekh Eksper.* 1964, 150.

Source and multiplier modifications to a time-of-flight mass spectrometer to improve sensitivity. Bitner, E. D., Rohwedder, W. K., and Selke, E., *App. Spectr.* **18,** 134 (1964).

Mass spectrometric studies of reactions in flames. I. Beam formation and mass dependence in sampling 1 atm. gases. Greene, F. T., Brewer, J., and Milne, T. A., *J. Chem. Phys.* **40,** 1488 (1964).

Observation and identification of ion dissociation processes occurring in the drift tube of a time of flight mass spectrometer. Hunt, W. W., Jr., Huffmann, R. E., and McGee, K. E. *Rev. Sci. Instr.* **35,** 82 (1964).

Time-of-flight mass spectrometer adapted for studying charge transfer, ion dissociation and photo-ionization. Hunt, W. W., Jr., Huffmann, R. E., Saari, J., Wassel, G., Betts, J. F., Paufve, E. H., Wyess, W., and Fluegge, R. A., *Rev. Sci. Instr.* **35,** 88 (1964).

Increased photographic sensitivity for time-resolved mass spectrometer data recording. Meyer, R. T., *Rev. Sci. Instr.* **35,** 1064 (1964).

Pulse generator system for fast-reaction experiments using a time of flight mass spectrometer. Riney, J. S. and Gardiner, W. C., Jr., *Rev. Sci. Instr.* **35**, 384 (1964).

Simple display system for recording time-of-flight mass spectra. Lincoln, K. A., *Rev. Sci. Instr.* **35**, 1688 (1964).

Synchronization of a flash lamp and mass spectrometer apparatus for fast reaction studies. Freese, J. M., and Meyer, R. T. (Scandia Corporation, New Mexico), SC-RR-65-185, Contract AT(29-1)-789 (1965).

Magnetic tape recording of analytical data. Issenberg, P., Bazinet, M. L., and Merritt, C., Jr., *Anal. Chem.* **37,** 1074 (1965).

A time-of-flight mass spectrometer. Matus, L., *Magyar Tud. Akad. Kozp. Fiz. Kut. Int. Kozlemen* **13,** 251 (1965).

Stationary film recording of time-resolved mass spectra. Moulton, D. McL., and Michael, J. V., *Rev. Sci. Instr.* **36**, 226 (1965).

Basic properties of electron multiplier ion detection and pulse counting methods in mass spectrometry. Dietz, L. A., *Rev. Sci. Instr.* **36,** 1763 (1965).

Rapid determination of electron impact ionization and appearance potentials. Martignoni, P., Morgan, R. L., and Cason, C., *Rev. Sci. Instr.* **36,** 1783 (1965).

Time-of-flight spectrometer for laser surface interaction studies. Bernal, E., Levine, L. P., and Ready, J. F., *Rev. Sci. Instr.* **37,** 938 (1966).

Time-of-flight mass spectrometers. Jayaram, R., *Mass Spectrometry*, Plenum Press, NY (1966), Chapter III.

Time-of-flight mass spectrometers. Blauth, E. W., *Dynamic Mass Spectrometers*, Elsevier (1966), Chapter IV.

Technique to improve signal to noise ratio of electron multiplier in a mass spectrometer. Young, J. R., *Rev. Sci. Instr.* **37,** 1414 (1966).

Computational methods for resolution of mass spectra. Luenberger, D. G., and Dennis, U. E., *Anal. Chem.* **38,** 715 (1966).

Computer assisted identification of mass spectrometric, flash filament signals. McCarroll, B., *Rev. Sci. Instr.* **38,** 444 (1967).

Shock tube–time of flight mass spectrometer apparatus with cryosorption pumping. Ryason, P. R., *Rev. Sci. Instr.* **38,** 607 (1967).

Method of operating an ion source for a time-of-flight mass spectrometer. Studier, M. H., US Patent 3,296,434 (1967).

Modification of a time-of-flight mass spectrometer for investigation of ion–molecule reactions at elevated pressures. Futrell, J. H., Tiernan, T. O., Abramson, F. P., and Miller, C. D., *Rev. Sci. Instr.* **39,** 340 (1968).

QUALITATIVE MASS SPECTRAL DATA

Application of mass spectrometry to structure problems. VI. Nucleosides. Biemann, K. and McCloskey, J., *J. Am. Chem. Soc.* **84,** 2005 (1962).

Fluorine compounds of xenon and radon. Chernick, C. L., Claassen, H. H., Fields, P. R., Hyman, H. H., Malm, J. G., Manning, W. M., Matheson, H. S., Quarterman, L. A., Schreiner, F., Selig, H. H., Sheft, I., Siegel, S., Sloth, E. N., Stein, L., Studier, M. H., Weeks, J. L., and Zirin, M. H., *Science* **138,** 136 (1962).

Volatile hop constituents: identification of methyl deca-4-enoate and methyl deca-4,8-dienoate. Buttery, R. G., Lundin, R. E., McFadden, W. H., Jahnsen, V. J., and Kealy, M. P., *Chem. and Ind.* 1963, 1981.

Further studies on the mass spectra of butanols: 3-monodeuterio-1-butanol and 4-monodeuterio-1-butanol. McFadden, W. H., Black, D. R., and Corse, J. W., *J. Phys. Chem.* **67,** 1517 (1963).

Volatiles from oranges. I. Hydrocarbons identified by infrared, NMR and mass spectra. Black, D. R., McFadden, W. H., Lundin, R. E., Schultz, T. H., and Teranishi, R., *J. Food. Sci.* **28,** 541 (1963).

Xenon tetroxides mass spectrum. Huston, J. L., Sloth, E. N., and Studier, M. H., *Science* **143,** 1161 (1964).

Mass spectrometric identification of decomposition products of polytetrafluoroethylene and polyfluoroethylene propylene. Kupel, R. E., Nolan, M., Keenan, R. G., Hite, M., and Scheel, L. D. *Anal. Chem.* **36,** 386 (1964).

Silicon–fluorine chemistry. I. Silicon difluoride and the perfluorosilones. Timms, P. L., Kent, R. A., Ehlert, T. C., and Margrave, J. L., *J. Am. Chem. Soc.* **87,** 2824 (1965).

Silicon–fluorine chemistry. II. Boron fluorides. Timms, P. L., Ehlert, T. C., Margrave, J. L., Brinckman, F. E., Farrar, T. C., and Coyle, T. D., *J. Am. Chem. Soc.* **87,** 3819 (1965).

Rearrangement of some piperidine *N*-oxides to hexahydro-1,2 oxazepines. Quinn, L. D., and Shelburne, F. A., *J. Org. Chem.* **30,** 3135 (1965).

Mass spectrometric studies of unsaturated methyl esters. Rohwedder, W. K., Mabrook, A. F., and Selke, E., *J. Phys. Chem.* **69,** 1711 (1965).

Specific rearrangements in the mass spectra of neopentyl esters. McFadden, W. H., Stevens, K. L., Meyerson, S., Karabatos, G. J., and Orzech, C. E., *J. Phys. Chem.* **69,** 1742 (1965).

Organic compounds in carbonaceous chondrites. Anders, E., Hoyatsu, R., and Studier, M. H., *Science* **149,** 1455 (1965).
Correlations and anomalies in mass spectra: lactoses. McFadden, W. H., Day, E. A., and Diamond, M. J., *Anal. Chem.* **37,** 89 (1965).
Correlations and anomalies in mass spectra: thioesters. McFadden, W. H., Seifert, R. M., and Wasserman, J., *Anal. Chem.* **37,** 560 (1695).
Doubly charged transition metal carbonyl ions. Winters, R. E., and Kiser, R. W., *J. Phys. Chem.* **70,** 1680 (1966).
Radiation chemistry of cyclopentane. Hughes, B. M., and Hanrahan, R. J., *J. Phys. Chem.* **69,** 2707 (1965).
Observation of astatine compounds by time-of-flight mass spectrometry. Appelman, E. H., Sloth, E. N., and Studier, M. H., *Inorg. Chem.* **5,** 766 (1966).
Feasibility of analysing ablator blow-off gases with a time-of-flight mass spectrometer. Lincoln, K. A., *NASA STAR*, **5,** 325 (1966).
Synthesis of thioborine. Kirk, R. W., and Timms, P. L., *Chem. Comm.* 1967, 18.
Boron–fluorine chemistry. I. BF and some derivatives. Timms, P. L., *J. Am. Chem. Soc.* **89,** 1629 (1967).
Chemistry of silicon difluoride. Margrave, J. L., and Thompson, J. C., *Science* **155,** 669 (1967).
Gaseous oxides and oxyacids of iodine and xenon: mass spectra. Studier, M. H., and Huston, J. L., *J. Phys. Chem.* **71,** 457 (1967).

GAS CHROMATOGRAPHIC EFFLUENT MONITORING

Time-of-flight mass spectrometry and gas–liquid partition chromatography. Gohlke, R. S., *Anal. Chem.* **31,** 535 (1959).
Some applications of mass spectrometry to lipid research. Dutton, H. J., *J. Am. Oil Chem. Soc.* **38,** 660 (1961).
Application of time-of-flight mass spectrometry and gas chromatography to reaction studies. Levy, E. J., Miller, E. D., and Beggs, W. S., *Anal. Chem.* **35,** 946 (1963).
Use of capillary gas chromatography with a time of flight mass spectrometer. Black, D. R., Day, J. C., McFadden, W. H., and Teranishi, R. *J. Food Sci.* **28,** 316 (1963).
Volatiles from strawberries. I. Mass spectral identification of the more volatile components. Black, D. R., Corse, J. W., McFadden, W. H., Morgan, A. I., and Teranishi, R., *J. Food Sci.* **28,** 478 (1963).
Fast-scan mass spectrometry with capillary gas–liquid chromatography in investigation of fruit volatiles. McFadden, W. H., and Teranishi, R., *Nature* **200,** 329 (1963).
Constituents of hop oil. Buttery, R. G., McFadden, W. H., Teranishi, R., Kealy, M. P., and Mon, T. R., Nature **200,** 435 (1963).
A rapid method for isolation and identification of sesquiterprene hydrocarbons in cold-pressed grapefruit oil. Hunter, G. L. K., and Brogden, W. B., Jr., *Anal. Chem.* **36,** 1122 (1964).

Volatiles from oranges. II. Constituents of juice identified by mass spectra. Corse, J. W., Kilpatrick, P. W., McFadden, W. H., Schultz, T. H., and Teranishi, R., *J. Food Sci.* **29**, 790 (1964).

Gas chromatographic separation of oxygen difluoride. Yee, D. Y., *J. Gas Chrom.* **3**, 314 (1965).

Cyclobutane compounds. III. The ionic addition of hydrogen chloride, hydrogen bromide and hydrogen iodide to allene and methylacetylene. Griesbaum, K., Naegele, W. and Wanless, G. G., *J. Am. Chem. Soc.* **87**, 3151 (1965).

Radiation chemistry of perfluorocyclohexane and perfluorocyclobutane. Fallgatter, M. B., and Hanrahan, R. J., *J. Phys. Chem.* **69**, 2059 (1965).

Identification of aldehydes and ketones from the time-of-flight mass spectra of their 2,4-dinitrophenylhydrazones. Light, J. F., and McKinney, R. W., *Appl. Spec.* **19**, 193 (1965).

Mass spectra of isocyanates. Ruth, J. M., and Philippe, R. J., *Anal. Chem.* **38**, 720 (1966).

Use of activated charcoal to trap gas chromatographic fractions for mass spectrometry and to introduce volatile compounds into a mass spectrometer. Damico, J. M., Sphon, J. A., and Wong, N. P., *Anal. Chem.* **39**, 1045 (1967).

Analysis of proteins, peptides and amino acids by pyrolysis–gas chromatography–time-of-flight mass spectrometry. Merritt, C., and Robertson, D. H., *J. Gas Chrom.* **5**, 96 (1967).

Multichannel open tabular columns. Merritt, C., Jr., and Walsh, J. T., *J. Gas Chrom.* **5**, 420 (1967).

REACTION MECHANISM STUDIES

Shock tubes

Shock wave studies by mass spectrometry. I. Thermal decomposition of nitrous oxide. Bradley, J. N., and Kistiakowsky, G. B., *J. Chem. Phys.* **35**, 256 (1961).

Shock wave studies by mass spectrometry. II. Polymerization and oxidation of acetylene. Bradley, J. N., and Kistiakowsky, G. B., *J. Chem. Phys.* **35**, 264 (1961).

Decomposition of nitro-methane in a shock tube. Bradley, J. N., *Trans. Farad. Soc.* **57**, 1750 (1961).

Mass spectral studies of kinetics behind shock waves. I. Thermal dissociation of chlorine. Diesen, R. W., and Felmlee, W. J., *J. Chem. Phys.* **39**, 2115 (1963).

Mass spectral studies of kinetics behind shock waves. II. Thermal decomposition of hydrazine. Diesen, R. W., *J. Chem. Phys.* **39**, 2121 (1963).

Observation and kinetic investigation of the NF and NF_2 radicals. Diesen, R. W., *J. Chem. Phys.* **41**, 3256 (1964).

Mechanism of chemi-ionization in hydrocarbon oxidations. Kistiakowsky, G. B., and Michael, J. V., *J. Chem. Phys.* **40**, 1447 (1964).

Shock wave studies by mass spectrometry. III. Description of apparatus; data on the oxidation of acetylene and of methane. Dove, J. E. and Moulton, D. McL., *Proc. Roy. Soc. (London)* A, **283**, 216 (1965).

Kinetics of the nitrous oxide decomposition by mass spectrometry. A study to evaluate gas-sampling methods behind reflected shock waves. Modica, A. P., *J. Phys. Chem.* **69,** 2111 (1965).

Mechanism of the acetylene–oxygen reaction in shock waves. Glass, G. P., Kistiakowsky, G. B., Michael, J. V., and Niki, H., *J. Chem. Phys.* **42,** 608 (1965).

Thermal decomposition of acetylene in shock waves. Gay, I. D., Kistiakowsky, G. B., Michael, J. V., and Niki, H., *J. Chem. Phys.* **43,** 1720 (1965).

Pyrolysis and oxidation of formaldehyde in shock waves. Gay, I. D., Glass, G. P., Kistiakowsky, G. B., and Niki, H., *J. Chem. Phys.* **43,** 4017 (1965).

Decomposition and oxidation of C_2F_4 behind shock waves. Modica, A. P., and Lagraffe, J. E., *J. Chem. Phys.* **43,** 3383 (1965).

Comment on NF_2 decomposition behind shock waves. Diesen, R. W., *J. Chem. Phys.* **45,** 759 (1966).

Pyrolysis of ethylene in shock waves. Gay, I. D., Kern, R. D., Kistiakowsky, G. B., and Niki, H., *J. Chem. Phys.* **45,** 2371 (1966).

Mass spectrometric and ultraviolet absorption study of CHF_3 decomposition behind shock waves. Modica, A. P., and Lagraffe, J. E., *J. Chem. Phys.* **44,** 3375 (1966).

Mass spectrometry studies behind shock waves. III. The thermal dissociation of fluorine. Diesen, R. W., *J. Chem. Phys.* **44,** 3662 (1966).

Mass spectrometric observation of ions formed during shock wave heating of gaseous Kr and Xe. Cresswell, R., DiValentin, M., and Dove, J. R., *Phys. Fluids* **9,** 2285 (1966).

Ethylene–oxygen reaction in shock waves. Gay, I. D., Glass, G. P., Kern, R. D., and Kistiakowsky, G. B., *J. Chem. Phys.* **47,** 313 (1967).

Kinetics of the difluoromethylene-NO reaction. I. Modica, A. P., *J. Chem. Phys.* **46,** 3663 (1967).

Kinetics of the reaction of fluorine with difluoroamino radicals and the dissociation of fluorine. Diesen, R. W., *J. Phys. Chem.* **72,** 108 (1968).

Flash photolysis

A mass spectrometric study of flash photochemical reactions. Kistiakowsky, G. B., and Kydd, P. H., *J. Am. Chem. Soc.* **79,** 4825 (1957).

Benzyne. Berry, R. S., Clardy, J., and Schafer, M. E., *J. Am. Chem. Soc.* **86,** 2738 (1964).

The dimerization of gaseous benzyne. Schafer, M. E., and Berry, R. S., *J. Am. Chem. Soc.* **87,** 4497 (1965).

1,4 Dehydrobenzene. A stable species. Berry, R. S., Clardy, J., and Schafer, M. E., *Tetrahedron Letters*, 1965, 1003.

Decomposition of benzenediazonium-3-carbonylate: transient 1,3 dehydrobenzene. Berry, R. S., Clardy, J., and Schafer, M. E., *Tetrahedron Letters*, 1965, 1011.

Dimerization of gaseous benzyne, Schafer, M. E., *Diss. Abs.* **27,** 444B (1966).

Flash photolysis and time-resolved mass spectrometry. I. Detection of the hydroxyl radical. Meyer, R. T., *J. Chem. Phys.* **46,** 967 (1967).

Reactions of excited iodine atoms. I. ($^2P_{1/2}$) in the flash photolysis of methyl iodide. Meyer, R. T., *J. Chem. Phys.* **46,** 4146 (1967).

Flash pyrolysis and laser evaporation

Products of the flash pyrolysis of phenol-formaldehyde by time-of-flight mass spectrometry. Friedman, H., *J. Appl. Pol. Sci.* **9,** 651 (1965).

Flash-vaporization of solid materials for mass spectrometry by intense thermal radiation. Lincoln, K. A., *Anal. Chem.* **37,** 541 (1965).

Flash pyrolysis of solid-fuel materials by thermal radiation. Lincoln, K. A., *Pyrodynamics* **2,** 133 (1965).

Mass spectrometer using a laser probe. Knox, B. E. and Vastola, F. J., *Chem. Eng. News* **44,** No. 44, 48 (1966).

Laser microprobe applied to mass spectrometry. Knox, B. E., and Vastola, F. J., *Laser Focus* **3,** 15 (1967).

Mass spectrometric analysis of carbon species generated by laser evaporation. Zavitsanos, P. D., General Electric, Missile and Space Division, USA, Tech. Inf. Series R67SD11 (1967).

Laser heating of coal particles in the source of a time-of-flight mass spectrometer. Joy, W. K., Ladner, W. R., and Pritchard, E., *Nature* **217,** 640 (1968).

Flames and molecular beams

Mass spectrometric detection of polymers in supersonic molecular beams. Greene, F. T., and Milne, T. A., *J. Chem. Phys.* **39,** 3150 (1963).

Composition profile of the diethyl ether/air two-stage reaction stabilized in a flat-flame burner. Agnew, J. T., and Agnew, W. G., *Tenth Internationa, Symposium on Combustion, Cambridge, England, 1964*, published 1965*l* 123 (*CA* **63,** 16204f).

Mass spectrometric studies of reactions in flames. II. Quantitative sampling of free radicals from one-atmosphere flames. Milne, T. A., and Greene, F. T., *J. Chem. Phys.* **44,** 2444 (1966).

Identification of flame ions by time of flight mass spectrometry. King, I. R., and Scheurich, J. T., *Rev. Sci. Instr.* **37,** 1219 (1966).

Mass spectrometric observations of argon clusters in nozzle beams. I. General behaviour and equilibrium dimer concentrations. Greene, F. T., and Milne, T. A., US Govt. Res. Rept., AD-637-177, **41,** N19, 74 (1966).

Mass spectrometric detection of dimers of nitric oxide and other polyatomic molecules. Milne, T. A., and Greene, F. T., *J. Chem. Phys.* **47,** 3668 (1967).

Miscellaneous

Photochemical oxidations. I. Ethyl iodide. Heicklen, J., and Johnston, H. S., *J. Am. Chem. Soc.* **84,** 4394 (1962).

Mass spectrometric investigation of the high temperature reaction between nickel and chlorine. McKinley, J. D., *J. Chem. Phys.* **40,** 120 (1964).

Mass spectrometric investigation of the nickel–bromine surface reaction. McKinley, J. D., *J. Chem. Phys.* **40,** 576 (1964).

Mass spectrometric investigation of the yttrium–chlorine surface reaction. McKinley, J. D., *J. Chem. Phys.* **41,** 2814 (1964).

Mass spectral study of the production of methylamine from azo-methane. Wacks, M. E., *J. Phys. Chem.* **68,** 2725 (1964).

Mass spectrometric study of desorption of oxygen from tungsten. Chuikov, B. A., Ptushinc'kii, and Yu, G., *Ukr. Fiz. Zh.* **9,** 1035 (1964) (CA **62,** 52f (1965)).

On the role of the nitroxyl molecule in the reaction of hydrogen atoms with nitric oxide. Kohout, F. C., and Lampe, F. W., *J. Am. Chem. Soc.* **87,** 5795 (1965).

Dynamic response measurements using a time of flight mass spectrometer. Douglas, J. M., *Chem. Eng. Sci.* **20,** 1142 (1965).

Mass spectrometric investigation of reactions of oxygen atoms with hydrogen and ammonia. Potter, E., and Wong, E. L., Lewis Research Centre, Cleveland, Ohio. OTS Document, NASA TND-2648 (1965).

Silicon–fluorine chemistry. IV. The reaction of silicon difluoride with aromatic compounds. Timms, P. L., Stump, D. D., Kent, R. A., and Margrave, J. L., *J. Am. Chem. Soc.* **88,** 940 (1966).

Mass spectrometric investigation of the nickel–fluorine surface reaction. McKinley, J. D., *J. Chem. Phys.* **45,** 1690 (1966).

Reactions of oxygen (^{3}P) atoms with formaldehyde. Niki, H., *J. Chem. Phys.* **45,** 2330 (1966).

Reaction of oxygen (^{3}P) atoms with diacetylene. Niki, H., and Weinstock, B., *J. Chem. Phys.* **45,** 3468 (1966).

Application of adiabatic compression apparatus for study of thermally induced homogeneous gas phase reactions. Barnes, V. M., Jr., Henderson, U. V., and Rhodes, H. A., *Rev. Sci. Instr.* **37,** 294 (1966).

Kinetic study on the reaction of oxygen atoms with dimethyl ether by time of flight mass spectrometry. Mori, S., Oishi, K., and Takeyaki, Y., *Bull. Inst. Chem. Res. Kyoto Univ.* **4,** 341 (1966).

Application of a time-of-flight mass spectrometer in studies of arc-quenching reactions in sulphur hexafluoride gas. Kamatani, A., and Miyamoto, T., *Elect. Eng. Japan* **84,** N4, 30 (1966).

Mass spectroscopic investigation of the reaction of oxygen atoms with methane. Wong, E. L., and Potter, A. E., *Can. J. Chem.* **45,** 367 (1967).

THERMAL ANALYSIS

Mass spectrometric thermal analysis (MTA). Langer, H. G., and Gohlke, R. S., *Anal. Chem.* **35,** 1301 (1963).

Obtaining the mass spectra of non-volatile or thermally unstable compounds. Gohlke, R. S., *Chem. and Ind.* 1963, 946.

Mass spectrometric techniques applied to the investigation of the thermal degradation of polymers. Wacks, M. E., *Conf. on High Temp. Polymers and Fluid Research*, ASD-TDR-62-372 (1962), 201.

Mass spectrometric studies of the thermal decomposition of poly (carbon monofluoride). Kuriakose, A. K., and Margrave, J. L., *Inorg. Chem.* **4,** 1639 (1965).

Thermal decomposition by mass spectrometry. Collin, J. E., and Delplace, A., *Bull. Soc. Chim. Belges* **75,** 281 and 304 (1966).

Thermal analysis by mass spectrometry. Gohlke, R. S., and Langer, H. G., *Fortsch. Chem. Fortsch.* **6,** 515 (1966).

KNUDSEN CELL

Uranium monosulphide. II. Mass spectrometric study of its vaporization. Cater, E. D., Rauh, E. G., and Thorn, R. J., *J. Chem. Phys.* **35**, 619 (1961).

Vapour pressures of scandium yttrium and lanthanum. Ackermann, R. J., and Rauh, E. G., *J. Chem. Phys.* **36,** 448 (1962).

Effect of oxygen on the vapour pressure of uranium. Ackermann, R. J., Rauh, E. G., and Thorn, R. J., *J. Chem. Phys.* **37,** 2693 (1962).

Application of time-of-flight mass spectrometry to the study of inorganic materials at elevated temperatures. Goldstein, H. W., Sommer, A., Walsh, P. N., and White, D. O., *Adv. in Mass Spec.* (published MacMillan, NY, 1963), **2,** 110.

Thermodynamic properties of actinide elements. Ackermann, R. J., and Thorn, R. J. *Thermodynamics of Nuclear Materials*, Proc. Symposium, International Atomic Energy Agency, Vienna, 1962, published 1963, 39.

Vaporization studies of thorium, uranium and plutonium metals and oxides. Ackermann, R. J., and Thorn, R. J., *Thermodynamics of Nuclear Materials*, Proc. Symposium, International Atomic Energy Agency, Vienna, 1962, published 1963, 445.

Vaporization of the rare earth oxides at elevated temperatures. Ames, L. L., Goldstein, H. W., Walsh, P. N., and White, D., *Thermodynamics of Nuclear Material*, Proc. Symposium, International Atomic Energy Agency, Vienna, 1962, published 1963, 417.

Vaporization studies on uranium–sulphur and uranium–oxygen–sulphur systems. Carter, E. D., Rauh, E. G., and Thorn, R. J., *Thermodynamics of Nuclear Materials*, Proc. Symposium, International Atomic Energy Agency, Vienna, 1962, published 1963, 487.

A thermodynamic study of the thorium oxygen system at high temperatures. Ackermann, R. J., Rauh, E. G., Thorn, R. J., and Cannon, M. C., *J. Phys. Chem.* **67,** 762 (1963).

A thermodynamic study of the tungsten oxygen system at high temperatures. Ackermann, R. J., and Rauh, E. G., *J. Phys. Chem.* **67,** 2596 (1963).

A literature review of mass spectrometer–thermochemical technique applicable to the analysis of vapour species over solid inorganic materials. Redman, J. D. (Oak Ridge Nat. Lab.), Report ORNL-TM-989 (1964).

Thermodynamic properties of gaseous yttrium monoxide. Correlation of bonding in Group III transition-metal monoxides. Ackermann, R. J., Rauh, E. G., and Thorn, R. J., *J. Chem. Phys.* **40,** 883 (1964).

Dissociation energies of diatomic molecules of the transition elements. I. Nickel. Kant, A., *J. Chem. Phys.* **41,** 1872 (1964).

The use of a pulsed time-of-flight mass spectrometer for investigating the evaporation of compounds. Mostovskii, A. A., and Sakseev, D. A., *Zhurnal Tekhn. Fiziki* **34,** 1321 (1964).

Dissociation energies of diatomic molecules of the transition elements. II. Titanium, chromium, manganese and cobalt. Kant, A., and Strauss, B., *J. Chem. Phys.* **41,** 3806 (1964).

Vaporization thermodynamics and dissociation energy of lanthanum monosulphide. Cater, E. D., Lee, T., Johnson, E., Rauh, E. G., and Eick, H., *J. Phys. Chem.* **69,** 2684 (1965).

Mass spectrum of Yttrium chloride vapour. McKinley, J. D., *J. Chem. Phys.* **42,** 2245 (1965).

Comments on pressure of uranium over uranium–uranium dioxide system. Ackermann, R. J., Rauh, E. G., and Thorn, R. J., *J. Chem. Phys.* **42,** 2630 (1965).

Time-of-flight mass spectrometric determination of dissociation energy of BeF. Hildebrand, D. L., and Murad, E., *J. Chem. Phys.* **44,** 1524 (1966).

High temperature equilibrium in germanium–nickel system. Dissociation energy of GeNi. Kant, A., *J. Chem. Phys.* **44,** 2451 (1966).

Atomization energies of the polymers of germanium. Kant, A., and Strauss, B. H., *J. Chem. Phys.* **45,** 822 (1966).

Dissociation energy of Cr_2. Kant, A., and Strauss, B., *J. Chem. Phys.* **45,** 3161 (1966).

Mass spectrometric studies at high temperature. IX. Sublimation pressure of copper (II) fluoride. Kent, R. A., McDonald, J. D., and Margrave, J. L. *J. Phys. Chem.* **70,** 874 (1966).

IONIZATION PHENOMENA AND APPEARANCE POTENTIAL MEASUREMENTS

Electron impact spectroscopy of ethylene sulphide and ethylenimine. Gallegos, E., and Kiser, R., *J. Phys. Chem.* **65,** 1177 (1961).

Studies of the shapes of ionization–efficiency curves of multiply charged monatomic ions. I. Instrumentation and relative electronic-transition probabilities for krypton and xenon ions. Kiser, R. W., *J. Chem. Phys.* **36,** 2964 (1962).

Electron impact spectroscopy of the four and five membered saturated heterocyclic compounds containing nitrogen, oxygen and sulphur. Gallegos, E., and Kiser, R. W., *J. Phys. Chem.* **66,** 136 (1962).

Electron impact spectroscopy of tetramethyl–germanium, trimethyl-silane and dimethyl-mercury. Hobrock, B., and Kiser, R. W., *J. Phys. Chem.* **66,** 155 (1962).

A technique for the rapid determination of ionization and appearance potentials. Kiser, R. W., and Gallegos, E. J., *J. Phys. Chem.* **66,** 947 (1962).

Ionization potentials of cyclopropyl radical and cyclopropyl cyanide (comunications to the Editor). Kiser, R. W., and Gallegos, E. J., *J. Phys. Chem.* **66,** 957 (1962).

The ionization potential of hydrogen disulphide (H_2S_2). Kiser, R. W., and Gallegos, E. J., *J. Phys. Chem.* **66,** 1214 (1962).

Electron impact spectroscopy of propylene sulphide. Hobrock, B., and Kiser, R. W., *J. Phys. Chem.* **66,** 1551 (1962).

Electron impact spectroscopy of sulphur compounds: 1,2,thiabutane, 2,thiapentane and 2,3,dithiabutane. Hobrock, B., and Kiser, R. W., *J. Phys. Chem.* **66,** 1648 (1962).

Electron impact spectroscopy of some substituted oxiranes. Wada, Y., and Kiser, R. W., *J. Phys. Chem.* **66,** 1652 (1962).

The ionization potential of (iso)thiocyanic acid. Shenkel, R., Hobrock, B., and Kiser, R. W., *J. Phys. Chem.* **66**, 2074 (1962).

Ionization and dissociation processes in phosphorus trichloride and diphosphorus tetrachloride. Sandoval, A., Moser, H., and Kiser, R. W., *J. Phys. Chem.* **67**, 124 (1963).

Electron impact investigation of sulphur compounds. II. 3-methyl-2-thiabutane, 4-thia-1-pentane and 3,4 dithiahexane. Hobrock, B., and Kiser, R. W., *J. Phys. Chem.* **67**, 648 (1963).

Electron impact investigations of sulphur compounds. III. 2 thiapropane and 2,3,4 trithiapentane. Hobrock, B., and Kiser, R. W., *J. Phys. Chem.* **67**, 1283 (1964).

Singly and doubly charged ions from methyl and ethyl isothyocyanates by electron impact. Hobrock, B., Shenkel, R., and Kiser, R. W., *J. Phys. Chem.* **67**, 1684 (1963).

Comparative ionization by million-electron-volt heavy ions and kiloelectron-volt electrons. Schuler, R. H., and Stuber, F. A., *J. Chem. Phys.* 1964, 2035 (erratum: *J. Chem. Phys.* **41**, 901 (1964)).

Appearance potentials of positive and negative ions by mass spectrometry. Melton, C. E., and Hamill, W. H., *J. Chem. Phys.* **41**, 546 (1964).

Ionization–efficiency curves for molecular and fragment ions for methane and the methyl halides. Tsuda, S., Melton, C. E., and Hamill, W. H., *J. Chem. Phys.* **41**, 689 (1964).

Appearance potentials by the retarding potential difference method for secondary ions produced by excited–neutral, excited ion–neutral, and ion–neutral reactions. Hamill, W. H., and Melton, C. E., *J. Chem. Phys.* **41**, 1469 (1964).

Structure of ionization–efficiency curves near threshold from alkanes and alkyl halides. Tsuda, S., and Hamill, W. H., *J. Chem. Phys.* **41**, 2713 (1964).

A study of threshold behaviour of electron impact induced ion and dissociation ionization of simple gaseous molecules. Glick, R. A., and Llewellyn, J. A., Bulletin No. 14, Division of Biology and Medicine, Inst. of Molecular Biophysics, USAEC (1964), Doc. No. FSU-2690-14.

Structure in the ionization efficiency curves of Ar_2^+ by pulsed mass spectrometry. Becker, P. M., and Lampe, F. W., *J. Am. Chem. Soc.* **86**, 5347 (1964).

Electron impact studies of some trihalomethanes. Hobrock, D., and Kiser, R. W., *J. Phys. Chem.* **68**, 575 (1964).

Mass spectral study of trimethyl borate. Wada, Y., and Kiser, R. W., *J. Phys. Chem.* **68**, 1588 (1964).

A mass spectral study of some alkyl substituted phosphines. Wada, Y., and Kiser, R. W., *J. Phys. Chem.* **68**, 2290 (1964).

Stabilization energies of substituted methyl cations—the effect of strong demand on the resonance order. Lampe, F. W., Martin, R. H., and Taft, R. W., *J. Am. Chem. Soc.* **87**, 2490 (1965).

An electron impact study of ionization and dissociation of trimethylsilanes. Hess, G. G., Lampe, F. W., and Sommer, L. H., *J. Am. Chem. Soc.* **87**, 5327 (1965).

Ionization potential and heat of formation of the difluoromethylene radical. Pottie, R. F., *J. Chem. Phys.* **42**, 2607 (1965).

Multiple ionization in neon, argon, krypton and xenon. Stuber, F. A., *J. Chem. Phys.* **42,** 2639 (1965).

Ions produced by electron impact with the dimetallic carbonyls of cobalt and manganese. Winters, R. E., and Kiser, R. W., *J. Phys. Chem.* **69,** 1618 (1965).

Ionization potentials and mass spectra of cyclopentadienyl-molybdenum dicarbonyl nitrosyl and 1,3-cyclohexadieneiron tricarbonyl. Winters, R. E., and Kiser, R. W., *J. Phys. Chem.* **69,** 3198 (1965).

Ionization and dissociation of ruthenium and osmium tetroxides. Dillard, J. G., and Kiser, R. W., *J. Phys. Chem.* **69,** 3893 (1965).

Mass spectrometric investigations of the synthesis stability and energetics of the low temperature oxygen fluorides. Malone, T. J., and McGee, H. A., Jr., *J. Phys. Chem.* **69,** 4338 (1965).

An electron-impact study of ionization and dissociation in methoxy- and halogen-substituted methanes. Martin, R. H., Lampe, F. W., and Taft, R. W., *J. Am. Chem. Soc.* **88,** 1353 (1966).

An electron impact investigation of some alkyl phosphate esters. Bafus, D. A., Gallegos, E. J., and Kiser, R. W., *J. Phys. Chem.* **70,** 2614 (1966).

Analysis of ionization efficiency curves and multiple ionization processes. Shadoff, L. A., *Diss. Abs.* **27,** 1445B (1966).

Mass spectrometric studies of the synthesis, energetics and cryogenic stability of the lower boron hydrides. Wilson, J. H., and McGee, H. A., *J. Chem. Phys.* **4, 6,** 1444 (1967).

Electron gun performance using a modified appearance potential technique. Sarzone, G., *NASA STAR* No. 22, 3982 (1966). *Mass Spec. Bull.* **1,** 781 (1967).

ION-MOLECULE REACTIONS

Charge transfer studies with a time-of-flight mass spectrometer: cross-section for N_2^+ on N_2 at 2650 eV. Hunt, W. W., Novock, R., and Stinnett, A. J., Tech. Rept. AFCRL-63-709 (1963).

Application of time-of-flight mass spectrometry to the examination of ion-molecule interactions. Homer, J. B., Lehrle, R. S., Robb, J. C., Takahashi, M., and Thomas, D. W., *Adv. in Mass Spec.*, Pergamon **2,** 503 (1963).

Specific reaction rate of the second-order formation of Ar_2^+. Lampe, F. W., and Hess, G. G., *J. Am. Chem. Soc.* **86,** 2952 (1964).

Ion–molecule reactions studied by time-of-flight mass spectrometry. Hand, C. W., and von Weyssenhoff, H., *Can. J. Chem.* **42,** 195 (1964).

Ion–molecule reactions studied by time-of-flight mass spectrometry. II. Reactions in CO-D_2 and CH_4-D_2 mixtures. Hand, C. W., and von Weyssenhoff, H., *Can. J. Chem.* **42,** 2385 (1964).

Ion dissociation in the drift tube of a time-of-flight mass spectrometer: spurious fragments arising from charge-transfer and dissociation reactions of retarded ions. Hunt, W. W., Jr., and McGee, K. E., *J. Chem. Phys.* **41,** 2709 (1964).

Gas-phase ion–molecule interactions involving atom transfer; limitations of the orbiting theory in accounting for the variation of cross-section with energy. Homer, J. H., Lehrle, R. S., Robb, J. C., and Thomas, D. W., *Nature* **202,** 795 (1964).

Observation of the products of ionic collision processes and ion decomposition in a linear, pulsed time-of-flight mass spectrometer. Ferguson, R. E., McCulloh, K. E., and Rosenstock, H. M., *J. Chem. Phys.* **42,** 100 (1965).

Mass spectrometric study of the bimolecular formation of diatomic argon ion. Becker, P. M., and Lampe, F. W., *J. Chem. Phys.* **42,** 3857 (1965).

Pulsed mass spectrometric study of the bimolecular formation of N_3^+. Cress, M. C., Becker, P. M., and Lampe, F. W., *J. Chem. Phys.* **44,** 2212 (1966).

Mass spectrometric study of ion–molecule reactions of ethanol and methanol. Sieck, L. W., Abramson, F. P., and Futrell, J. H., *J. Chem. Phys.* **45,** 2859 (1966).

High transmissions and dual electron beam ion sources for mass spectrometry. Melton, C. E., *J. Sci. Instr.* **43,** 927 (1966).

NEGATIVE IONS

Autodetachment of electrons in sulphur hexafluoride. Edelson, D., Griffiths, J. E., and McAfee, K. B., Jr., *J. Chem. Phys.* **37,** 917 (1962).

Non-dissociative electron capture in complex molecules and negative ion lifetimes. Compton, R. N., Christophorou, L. G., Hurst, G. S., and Reinhart, P. W., *J. Chem. Phys.* **45,** 4634 (1966).

MISCELLANEOUS

Measurement of breathing by analysis of respiratory gases by time-of-flight mass spectrometry. Caldwell, P. R. B., and Graves, J. B. (Aerospace Med. Div., Wright Patterson Airforce Base, Ohio), US Govt. Reprt. (1965), AD 608339 (AMRL-TR-64-84).

Observations on rats exposed to a space cabin atmosphere for two weeks. Felic, P., *Aerospace Medicine*, **36,** 858 (1965).

Direct observation of the decomposition of multiply charged ions into singly charged fragments. McCulloh, K. E., Sharp, T. E., and Rosenstock, H. M., *J. Chem. Phys.* **42,** 3501 (1965).

Comparative stabilities of gaseous alane, gallane and indane. Breisacher, P., and Siegel, B., *J. Am. Chem. Soc.* **87,** 4255 (1965).

Basic studies in radiation technology. Final report, JLI-2901-75, USAEC, Johnston Labs. Inc., Baltimore (1966).

Low energy investigation of radiation chemistry. Kiser, R. W., *Mass Spec. Bull.* **1,** 346 (1967).

AUTHOR INDEX

SUBJECT INDEX

I. TIME OF FLIGHT MASS SPECTROMETER

II. GENERAL INDEX